EPIC
MEASURES

EPIC
MEASURES

One Doctor. Seven Billion Patients.

JEREMY N. SMITH

HARPER WAVE

An Imprint of HarperCollins*Publishers*

HarperCollins books may be purchased for educational, business, or sales promotional use. For information, please e-mail the Special Markets Department at SPsales@harpercollins.com.

FIRST EDITION

Designed by William Ruoto

Image on page iii by Kevin Van Aelst

Library of Congress Cataloging-in-Publication Data has been applied for.

ISBN: 978-0-06-223750-7

15 16 17 18 19   OV/RRD   10 9 8 7 6 5 4 3 2 1

To Crissie McMullan and Jane Smith
&
To John Benson and the study group

He who conceals his disease cannot expect to be cured.

—ETHIOPIAN PROVERB

CONTENTS

PART III: RESURRECTION

PART IV: GOING LIVE

Counting Everything When Everything Counts

What you don't know can kill you—A genius and a madman—The human side of scientific revolutions.

We are told we live in the age of Big Data. From hedge funds to Internet search algorithms to baseball sabermetrics, numerical analysis—on an unprecedented scale—guides more and more of our decisions. As I write, you can pay $99 for a "personalized genome service"—23andme—that uses a saliva sample to provide one million points of data from your DNA, to tell you about your ancestry and warn you about your propensity for certain diseases (though the health warnings have been suspended by directive of the United States Food and Drug Administration). Another $99 and you can buy a wearable device like the Fitbit, which tracks your every move—even how well you sleep.

But basic information about what actually kills people and makes them sick is trickier to tabulate. In 2010, approximately 53 million people died worldwide, and, for all but a fraction, no one knows definitively why. In 147 of 192 countries, reliable death certificates—often any death certificates—don't exist, and, even in rich nations, health records have many missing pieces. Consider these basic questions: In the United States, one of the wealthiest countries in the world, does life expectancy vary depending on where you live? How different are the causes of illness and injury for men and women? Do Americans spend more time suffering from job-related accidents or outdoor air pollution, from drug abuse or not eating enough fruit? Incredibly, no one has really known. And yet efforts to help everyone in danger are stymied if we don't know who is getting sick and dying, and why.

Health, to date, has generally been counted in two crude ways: length of life and cause of death. These measures are very poor reflections of how we all actually live—mere epitaphs, not biographies. If you are anemic, arthritic, deprived of sight, or depressed, you are very far from perfect health, but you may live just as long as other people, and something else will likely kill you. That no one dies from a migraine doesn't mean headaches don't have consequences. That there are no pink ribbons for low back pain doesn't mean it doesn't hurt and cost days at work. Chronic conditions like these drive a huge and growing proportion of private and public health spending—and, of course, of human suffering. If we want to improve how we live as well as how we die, we need to know the full measure of our diseases and disabilities—what doesn't kill us as well as what does.

Ignorance is expensive. Between 1990 and 2010, international development assistance for health—medical aid money—more than quintupled from $5.8 billion to $29.4 billion a year. And that's minuscule compared with what countries and individuals spend on themselves. At last count, annual total health spending worldwide was $7 *trillion*—10 percent of the global economy and growing. But is that money being spent on the health threats that really cause the most suffering, or only on what *seem* to be our worst problems? Are billions of lives at risk and trillions of dollars being wasted because of priorities based on faulty information?

Everyone wants the world to move in a healthier direction. But what we need is a map. And if no accurate, sufficiently extensive map exists, someone needs to create one.

This book is the story of a huge independent effort, years in preparation, to do nothing less than chart everything that threatens the health of everyone on Earth, and make that information publicly available to doctors, health officials, political leaders, and private citizens everywhere. The quest has engrossed the time and talent of thousands of people around the world, from computer programmers to village interviewers. Chris Murray, the originator and now leader of the project, has been called a genius and a madman: a Harvard-trained physician who no longer practices medicine but is trying to treat the world's 7 billion people, an Oxford-educated economist who doesn't follow the stock market but is believed by some to hold the key to one of the largest segments of the international economy. You might also say he is a very smart guy who has found a way to channel an obsession with detail, a prodigious appetite for hard work, and an unusual kind of global compassion into the monumental task of surveying, comparing, and combating all the illness and injury, fatal and disabling, that burdens each and every human being. That is the study's name, in fact: the "Global Burden of Disease."

Global Burden is a concept, a quantity, and an ongoing project—a comprehensive, comparable measure of almost everything wrong with everyone everywhere. Its numbers can be broken down by person, place, ailment, and consequence—what kills us, what makes us sick, and what shortens our pain-free years of life. It can identify the probable top killers of newborn children in Angola or of middle-aged men in the United States, the worst causes of pain and suffering for teenagers in Egypt or for elderly

French women, and the global toll of everything from asthma to suicide to chronic neck pain. It is not a static document, but an evolving report, in ever greater detail, that has already released a trove of more than 650 million results. These may provide more powerful ammunition in the fight against unnecessary suffering and needless death than any other invention in the history of public health. The basic principles of a medical practitioner apply to the 7 billion as well as to the individual patient. First, diagnose. Then prescribe.

What are the world's health problems? Who do they hurt? How much? Where? Why? Forget what you think you know. With a truly all-encompassing view of life and death, we can see for the first time if Europe is healthier than America, or Iowa than Ohio, or you than your neighbor. And then in what ways. And how people are responding, with specific details everyone else around the world can try to emulate.

The question then becomes not what stops us from living better, but how far and how fast are we willing to improve?

first met Chris Murray in January 2012. The project he described was one of the largest scientific exercises ever attempted. It was as complex and controversial as the first moon landing or the Human Genome Project. It was extremely expensive, insanely ambitious—and almost done.

Murray himself was fascinating: blunt, often abrasive, hyperenergetic, supremely confident, yet fiercely collaborative. As his colleagues would all attest, he has always been a person who likes to argue, and he seemed to operate on the assumption that scientific progress relies on picking fights. He was also intellectually generous, invigorated by the push and pull of other people's ideas and willing to listen to any serious proposition, no matter the source.

Soon that list of outside ideas included the notion that I be allowed to watch as he and his team scrambled to complete the latest, most significant stage of an effort that had started more than twenty years before.

Murray agreed. He set no restrictions on my questions, whom I talked to, or what I saw, and had no control over what I wrote. This was brave, perhaps even reckless—he had made prominent enemies, he had personal secrets, and his project might well fail—but it was also in character. The longer I observed Murray, the more the question of personality interested me. Before meeting him I had considered myself a lively person of above-average stamina. Follow him for just twenty-four hours, though, and I required a week afterward to recover. For fun, he raced sailboats, skied on virgin slopes reached via helicopter, and mountain-biked across forest and desert. He was at once personally reserved and, categorically, an extrovert: "Basically, I am only capable of creative thought when I am interacting with others," he said to me. But if he was convinced you were wrong, he paid no attention to what you said, no matter who you were or how exalted your position. "Do good work that matters" was one of his personal mottos. Another was "Everything everyone tells me is a lie until I can verify it's the truth."

We can't wait for a better map of what's ailing us, Murray said—and we don't have to. New methods of analysis and new powers of computation make it possible to unite previously scattered points of information in revelatory ways. One use of the discipline of Big Data, much chronicled by the media, is taking almost-infinite stores of knowledge and reducing them to a single answer (think Google). Another, relatively neglected by reporters, is taking extremely sparse data and ingeniously stitching them together to construct a provably reliable big picture. A third is finding and correcting errors in the information we already have. Murray's claim was to have mastered all of the above in service of the most essential question of all: how to measure—and improve—how we live and die. And everyone, everywhere was included—now and for all time.

This was a tall order, but, when Murray made the case, the impossible seemed not only possible but necessary. It's not acceptable, he said, not to know what people die from around the world. It's not acceptable to count only rich countries or only causes that have a spokesperson. It's not acceptable to ignore nonfatal conditions or to let the powers that be decide what's important without outside oversight or public input. And it's simply shortsighted to take just what we already know and then see what it tells us. Instead we have to decide what we *need* to know and then go out and get that information.

This is what Murray and his colleagues have done, and continue to do. If you have ever read that the U.S. health system is ranked 37th in the world, that famous (to some, infamous) figure comes from their studies. Whether identifying tuberculosis as the leading infectious killer of adults at a time when most global health programs focused only on diseases of children, or revealing in which U.S. counties men and women live longer than their counterparts in Japan (and in which they die earlier than in Syria), their work makes headlines and resets the priorities of national and international health organizations. They have shown the wealthiest couple on Earth a way to invest their fortune for global good. And they may help any of us, anywhere, know what really hurts us and what will best improve our health.

The people transforming our knowledge of life and death are not saints. They are very much human beings, albeit extraordinary ones. They boast human virtues and they suffer human flaws. Saying that the way we measure health is broken and that you can fix it requires a conviction, drive, and focus that almost all of us would find inconceivable. It means making enemies of good people who stand in your way or who you believe are wrong or wrongheaded. It means overcoming politics and embracing competition—for money, for power, for priority.

How the Global Burden of Disease study came into being— and what it can tell us already—is an epic tale. It encompasses wars and famines, presidents and activists, billionaires and billions of people worldwide living in poverty. It shows the human side of scientific revolutions—and of revolutionary scientists: their mistakes and setbacks as they happen, their personal foibles and frustrations, how they face critics and rivals, and if and how they can ever claim success.

But even revolutions have small beginnings. This one started more than forty years ago, in a Land Rover crossing the Sahara Desert.

PART I

Who Dies of What

CHAPTER ONE

Murray, Murray, Murray, and Murray

The navigator—A childhood memorizing maps—"Do you have some water?"—*Medicament*—A deadly puzzle—Both skeptic and true believer.

March 1973. The Sahara.

There was no road, and certainly no GPS. Forward motion meant following a dusty track. Occasionally, through the haze, the family had seen a lone gazelle or a few people on camels. Every now and then they had discovered a village. For the last three days, however, they had encountered no one but themselves. Drought and daytime temperatures touching 120 degrees Fahrenheit made the area almost uninhabitable. Now, at four in the afternoon, they came to a split in the track and didn't know which way to go.

John, hair white, his bald spot susceptible to sunburn, wearing professorial black-framed glasses, drove one dark green Land Rover. Anne, an athletic redhead, accompanied him or Nigel, their seventeen-year-old son, who drove the other. Luggage, tents, bedding, food, a cookstove, and other supplies for the trip lined every free inch of the vehicles' interiors. In the backseats, Megan, fourteen years old, and Christopher, ten, couldn't touch their feet to floors packed with flat-sided five-gallon metal jerry cans. The ones with water were still full, though sloshing, the children heard. The ones with gas, already partly empty, seemed to make a louder sound. Let them accidentally clank against each other and they might echo ominously.

The adults conferred about which trail to follow. Megan, who wanted to be an anthropologist, passed the time imagining what it would be like to live where they were going. Chris, the family navigator, brushed overgrown brown bangs from his eyes and studied yet again the only map of their terrain.

Drawn by French surveyors at a scale of 1 centimeter to 40 kilometers, the map depicted the desert in mustard yellow. Every morning and evening, before breaking camp or bedding down for the night, the boy ran to lead his father and older brother in unfolding it for reference atop the hood of a Land Rover. Tiny symbols—an X, an empty square, a stick-figure house—marked possible stops for gas, repairs, and primitive lodging. "Eau bonne à 5 m"—good water at 5 (or 15 or 35) meters—little notes suggested along the route. Of the unmarked trails they followed, Chris read: "Suitable only for cross-country vehicles and certain types of truck. To use them a guide or means of land navigation is necessary. Traveling with only one car inadvisable."

Chris, dressed in a short-sleeve collared shirt and shorts, was noticeably thin, all knees and elbows; at home, in Golden Valley, Minnesota, his parents had tried to fatten him up with eggnog and ice cream. His energy and diligence, however, made his presence larger. "Knowing where you were was a big thing," he would say years later. "Crossing the Sahara, it's a matter of life and death." Now, carefully double-checking their whereabouts, he calculated that their next fuel stop was not for five hundred kilometers.

The family decided to drive left. Sweating, they spent an hour traversing rough chunks of rock, down into a valley or gorge. Then dunes, nearly liquid in the heat. Standing at the edge of an escarpment, the Murrays could see no more track below them. They went back, exploring the other side. Finally, the path was pure loose sand. Wrangling "sand ladders," six-by-two-foot metal boards, full of holes, under buried tires, they returned to the original split.

In England, John, the intellectual, had bought a compass. Now Chris berated him for not using it. Instead, as the sun set, they prepared to wait as long as necessary for a fellow traveler of whom they could ask directions. A guide—which was supposed to be Chris's job. "He made some pretty nasty comments," John defended himself decades later. "But we didn't know what the sand was like. If we had gone off, we could have been lost in it."

This was how the Murray family arrived in Africa. The hard part was yet to come.

Chris Murray spent his entire childhood memorizing maps. His parents were both New Zealanders—the most travel-mad people on Earth. John was a cardiologist. Anne was a microbiologist. They had even met in motion—as students in 1943 on a train back to the University of Otago following a school break. In the 1950s, a stint together at the Mayo Clinic and then the offer of a professorship for John at the University of Minnesota brought them to the United States.

Exploration became the family passion. Winters, John and Anne packed Chris and his three older siblings—Linda, Nigel, and Megan—on car trips to the new ski resort of Vail, Colorado. Summers, they drove to southern California, where they camped on the beach. To see as much country as possible, the Murrays traveled one summer through Yellowstone and the Tetons; another summer to Oregon and down the Pacific coast; and a third year through Colorado. To save money en route, John drove late into the night, stopping the car by the side of the road where he set up cots for everyone to sleep on. Anne, who had grown up on a remote dairy farm, taught the children to embrace the adventure of unfamiliar places. "She wanted to see over the next hill and around the next bend—always," John says.

By the mid-1960s, Linda, Chris's eldest sibling, had graduated from college and started work as a flight attendant for Pan American Airlines. A job benefit allowed family members to fly on standby anywhere in the world for only 20 percent of the normal fare. Suitcases packed, sleeping in airports if necessary, the Murrays took off every chance they could, visiting Thailand, Turkey, Lebanon, Egypt, and India. Once the family flew to Nairobi, rented a minivan camper, and spent a month touring Kenya, Uganda, and Tanzania in a big circuit. Then, inspired by seeing the Omar Sharif movie *The Horseman*, Anne decided they should go to Afghanistan. More than four decades later, Chris, the baby, would still recall the blue lakes of Band-e Amir, the 120-foot-plus sandstone Buddhas of Bamiyan (later destroyed by the Taliban), and a giant pile of skulls he was told were remnants of Genghis Khan's rampage through the region seven and a half centuries earlier.

In 1973, John was granted a sabbatical for the next academic year. He suggested that the Murrays spend it in South Africa, home of the cardiac surgeon Dr. Christiaan Barnard, who had performed the first successful human heart transplant. Nigel, now a high school senior, refused. Emphatically. Even though his father meant to spend the year in basic research, not politics, the long-haired teenager would not live in a country ruled by apartheid. Megan, a high school freshman, and Chris, a fourth grader, agreed. If the family had an entire year free ahead of them, they should serve people in need directly, the children said. Make themselves useful. Weren't there millions around the world desperate for medical attention?

Anne, inspired, started planning. "She just loved the desert and everything about it," John says of their earlier trips. "The radical change"—from being a farm girl, then lab worker, and, finally, stay-at-home mother in suburban Minneapolis—"to the life of adventure." Through their local church, Westminster Presbyterian, Anne contacted people at Church World Service. They said a new hospital had been built in sub-Saharan Africa, in eastern Niger, then and now one of the poorest countries on Earth. Could the Murrays help there?

The family sat at the kitchen table, three children and his wife lobbying John. "That's a good idea," Nigel, Megan, and Chris said. "Why don't we do that?'"

"We can make a contribution to humanity," Anne said. "A small one. Nothing grand. Work together as a family, do something collectively."

Fine, John said. Maybe the challenge would be good for them. His father had come to New Zealand from Lebanon at the age of five or six, and had gone door to door selling matches for pennies. Neither of his parents had graduated from high school. Cabaret-style restaurateurs, they paid for his medical education by singing duets as they sold roast dinners to American GIs. It frustrated him that none of his kids, growing up in prosperous suburbia, seemed particularly motivated students. They wanted to spend a year working in Niger? If anything can change your life dramatically, he thought, this is it.

The family raised money, flew to England, and stayed in Oxford while gathering supplies for the year ahead. They bought the Land Rovers on discount direct from the factory in Solihull. They went by ferry from Southampton to France, drove south into Spain, and crossed over into Africa from near Gibraltar.

Then it was into the world's largest desert.

The Sahara covers an area of some 9.4 million square kilometers, almost equal to Europe, or more than thirteen Texases. In three directions, water determines its borders: the Mediterranean to the north, the Atlantic to the west, and the Red Sea to the east. Southward, the desert ebbs into the Sahel, a semiarid region long known for greater moisture, and is therefore much more populated. But as John, Anne, Nigel, Megan, and Chris drove slowly south hundreds of kilometers to Niger's capital, Niamey, and then east another 1,300 kilometers to their assigned hospital, in the town of Diffa, near Niger's borders with Chad and Nigeria, the heat and drought never abated. People would wave down their Land Rovers. They were begging. At first, the Murrays thought they wanted money. Well, money was useless. "Do you have some water?" the family was asked instead. "Can I have some water for my baby?"

They arrived in Diffa, a district capital, in early April. One to two thousand people lived here, most in straw-roofed mud huts. The hospital, really a little clinic, donated by Italians, was easy to spot: two long, low one-story prefab buildings, the only modern structures in a very unmodern town. One building, outpatient services, held a waiting area, examining rooms, a lab, and a pharmacy. The other building, a hospital ward, had ten beds each for men and for women, plus nurses' quarters, a surgical suite, and an operating room. All were empty.

There was housing for an electrical generator that had not yet come. A water tower with no water. A supply room with almost no supplies.

Outside people gathered in the sand. The patients.

The Murrays asked to meet other staff: doctors, nurses, administrators. A government official appeared. It turned out that another doctor who had arrived previously had taken one look at the setup and left.

The family was furious. "It's no good to us," John said. "We can't operate in this situation." Or could they? Brought all the way from Minneapolis and bolted to the Land Rovers were their own generator and portable electric cardiograph machine, an old microscope and basic stains for lab work, and medical supplies to last perhaps two weeks. He conferred with Anne a world away from their comfortable kitchen table. She'd wanted an adventure? Now she had one. If they stayed, John would be the chief—and only—physician. Anne and Megan, the fourteen-year-old, would be nurses. Nigel, the seventeen-year-old, would fix equipment and man the lab. And Chris, the ten-year-old? "I was the pharmacist and run-around boy," he explains.

Or they could go back all the way they'd come.

"We asked ourselves, can we operate this hospital without water and without electricity?" John remembers. "We decided we could."

A little doctor's house sat beside the clinic, surrounded by thin bushes and pink-and-white flowering portulaca. Unfortunately, it had been built without windows, only air conditioners. Like the clinic itself, these probably appeared as a great advance for Africa on the balance sheets of Italian aid. Without electricity, however, they were a worthless encumbrance. Except when it rained, which was very seldom (the Sahelian drought was the worst in twenty years), the family camped outside. They threw up sheets for privacy, dropped mosquito nets, and unfolded little cots they folded up again and put away come morning. Waking up, they saw the bed nets above them. They immediately put on boots, checking first for scorpions.

Centuries ago, Lake Chad covered the entire area, but the drought and dry climate had shrunk this shallow water body to a fraction of its former size, pushing the closest access a torturous hundred-kilometer trek east. A single well in the town square, a thousand feet deep, sustained everyone. Local food, in short supply, was the past year's millet, ground to powder. Every morning, Chris heard a rhythmic pounding as women mixed the millet with spices and water to the consistency of porridge.

People heard the hospital had opened. And a doctor was offering medical care. They walked days and nights to have themselves, their children, or their elderly relatives treated. John went to the local prefect and asked permission to use the well. This agreed to, Nigel was sent over every morning with the Land Rovers. Laboriously, he filled the family's jerry cans and a donated 200-liter drum. Then, since the Murrays lacked working radios or an ambulance service, the teenager roamed the area, distributing water as necessary and picking up people too ill to make the trip on their own. Megan, meanwhile, greeted patients in morning rounds with her mother. As directed by her father, she dispensed medication, IVs, shots, or stitches. Every day, they had to sweep the floors of sand and dust. Every three days, they swabbed them. As far as electricity was concerned, the family was helpless. What little fuel they had went to run the generator during key procedures. Otherwise, cool came from shade and nightfall, and light from the sun. "If we finished what we were doing in the hospital, we'd attend my dad," Megan says. They held flashlights while he was doing routine surgery.

Chris, too young at first to provide direct care, briefly attended the local school. The experience was a disaster. Bookish and bright, more like his academic father than any other sibling,

Chris loved brainteasers and playing board games like Risk. Neither prepared him for the strictures of "learning" in a one-room hut. Nobody spoke any language he knew. Any student who did anything wrong got beaten. He contracted hepatitis A, a fever-causing virus transmitted in food, and lost almost 40 percent of his already-meager body weight, dropping from 89 to as few as 54 pounds before stabilizing. "Chris suffered terribly," says John. "We were desperate." They drove 725 kilometers south and west to Kano, a major city in northern Nigeria, for supplies and fresh food. Chris's skin was bright yellow. It took all his effort, he would remember later, not to fall down outside the main hotel.

When he recovered, his parents gave Chris the job of organizing the clinic's precious few medical supplies. In the dry, dusty-smelling room, he witnessed his father make frantic phone calls, requesting necessary equipment and basic drugs such as penicillin. A shipment was on its way, they were told. A paper voucher followed: *medicament*, it read. French for "medicine." Wonderful. Chris and his family waited eagerly for weeks. At last a full truck lurched across the horizon.

They pulled down the flaps. There were no drugs. What filled the pickup instead were hundreds, even thousands, of cans of marmalade. All of them had gone bad somewhere. It was ludicrous—"a moment from the theater of the absurd," says Chris. "You had to cry and laugh at the same time," says Nigel. "How did we get from *medicament* to marmalade?" Was it a mistranslation? Had somebody cynically wanted to get rid of rotten food? Was it outright fraud? And did anybody in power care, or did they just want to check a box that said the Murrays had been answered? There was no way of knowing. But two things were clear to Chris. Just because somebody in authority promised something didn't mean it was going to happen. And the only way to know true from false was to find out for yourself.

Chris started helping out in the wards. He saw anthrax, tuberculosis, a viper-bit gangrenous leg, the mouth of a guinea worm popping out from an ankle. "At the time, it was just part of life," he would say later. "This was just what we were doing. It was early enough in my childhood, I never thought it was something kids shouldn't do." One day around lunchtime, Chris found an old man outside the hospital, prostrate in the sand by the facility's empty water barrels. "He was very proud," Chris says. "He didn't want to show people he was sick." But Chris was a child. "He dragged me over and showed me where he'd vomited blood in the sand." The ten-year-old ran to get his father and show him what he'd seen. He held the man's hand as the Murrays transported their patient back to the hospital.

John diagnosed a complication of cirrhosis, caused not by alcohol but by schistosomiasis—parasitic worms invading varicose veins to cause a massive hemorrhage from the esophagus. The man was so grateful for Chris's help, he had his relatives give the boy the gift of a bag of limes, an almost unheard-of luxury in the area. A few days later, however, in bed, the man's body exploded in blood. Another hemorrhage, this one fatal. Chris was very upset. But he returned almost immediately to work.

Morning to night, stricken families carried sick children his own age or younger in from the desert wrapped up in brightly colored cloth. These children looked incredibly, horribly skinny, nothing but rags and protruding rib cages. Some screamed. Others lacked even that energy. Many died. Worst was a desperately malnourished child carried all day to the hospital in a basin of water. The parents' intention was to stop fever. When they lifted the towel over the basin to show the Murrays, however, their tiny child was dead of drowning.

Chris's older siblings were deeply troubled by what they saw. "You can understand death in adults," Nigel says. "In kids—

you're still a kid yourself at that age—it's your own kind." Seeing so many perish, he remembers, sent "shock waves" through his psyche. Forty years later, Megan still choked up at the memory. Chris, though, tried to emulate the stiff upper lip of his parents. It would be a lifelong pattern in him: repressing negative emotions and channeling his energy instead on work. Doctors can't cry—and certainly not in front of their patients. They have to focus to save lives. "His personality became much more outgoing," John says. "He became much more assertive. A little more obsessive."

Step one of treating dehydrated and malnourished patients was attempting to get fluids into the sufferers. Step two was going to the market and buying food to try to feed them. As the weeks passed, however, the Murrays noticed something first peculiar, then disturbing. Despite the drought and famine, the tribespeople around them were almost entirely free of malaria and common viral illnesses. This freedom seemed to contradict the basic tenets of nutritional science, which said a starved body would soon get sick. What was disturbing was that the situation reversed once people were admitted to the hospital. There malaria was endemic, whatever someone's primary reason for admission. You didn't even have to be sick in the first place to suffer the disease. Healthy visiting relatives became victims, too.

It was baffling, particularly in the driest of dry seasons, when neither rain nor mosquitos could be blamed. Something at the hospital itself was bringing everyone—patients or not—close to death. But what? "Maybe we're poisoning them with the vitamin pills we're handing out," Megan said.

She was joking, but the one thing given both patients and relatives from out of town was food. This made sense. They were *starving*. Adult women averaged 96 pounds. Adult men averaged 112 pounds. How could nutritional supplements or extra calories be bad for them?

Yet John Murray was the kind of academic who encouraged his children to be skeptical of conventional wisdom. Throughout his career as a medical researcher, he had studied the metabolism of iron, which affects all sorts of ailments from preterm birth to heart disease. The mainstream scientific view was that iron deficiency should always be avoided. And maybe so. Even in the United States, children were said to be iron-deficient, and drug companies loaded kids' vitamins with the metal. But what if certain types of parasites, key to infectious diseases, *also* thrived on iron? As the family fed patients and their relatives, John theorized, they might simultaneously be nurturing those parasites. They couldn't let the people starve, of course, but indiscriminate feeding might be as bad or worse. "When you have a situation that arises that's at variance with general observation," John said, "you've really got to stand back and look at it dispassionately." It was time to find the facts.

To test whether the food they provided was somehow spurring malaria, John had Chris and Megan record the height, weight, and general nutritional state of adult patients and accompanying relatives over the age of sixteen, first on everyone's arrival at the hospital, again after forty-eight hours, and then a final time five days later. At the same intervals, Anne and Nigel drove up a Land Rover, hooked their microscope to its generator, and measured hemoglobin levels, red blood cell counts, and serum iron and total iron-binding capacity.

No one tested had observable malaria before entering the facility. All were given dried skim milk, grains, and multivitamins without iron. Subsequent attacks struck 23 of 72 patients and 51 of 109 relatives—two in every five people fed. They responded to the antimalaria drug chloroquine, but what had made them sick?

Could the Murrays really blame iron? Peak frequency of malaria was day five after arrival. As the lab tests showed, day five was also when iron in the blood, after reaching maximum saturation, began to fall. It certainly seemed suspicious—like fresh footprints leading from a murder victim to the house of his or her best friend. How, though, to spread the warning to the wider world? And once they did, would anyone believe them?

Chris returned with his parents to Minnesota. But their experience in Africa had raised questions that the Murrays still needed to answer. They decided to write up what they'd observed about the mysterious presence of malaria in their clinic—not as an opinion piece or memoir, but as a scientific paper. To the family's clinical observations in the Sahel, John added an experiment of his own design, giving rats with malaria intramuscular iron, which turned out to speed infection. Anne, meanwhile, did a historical review of previous studies relevant to iron-deficiency treatment and the onset of disease. Not bothering with baby steps, they submitted the paper to the British medical journal *The Lancet*. Since his children were so involved in the study, John decided to put Nigel's and Megan's names on it. And that's how the editors accepted it. "Refeeding-Malaria and Hyperferraemia," the first paper authored by Murray, Murray, Murray, and Murray, appeared on March 22, 1975. Now their findings would reach a larger community of doctors and public health workers.

The Lancet, founded in London in 1823, is one of the most influential and prestigious scientific publications in the world. What appears in it is read not only by other leading scientists, but also by policy makers and the press worldwide. After the article appeared, the BBC called John. They asked, "Do you believe in depriving everybody of food?" He laughed. "Of course not." He just didn't want to feed people if it would make their lives worse. No one had died from the observational experiments in Diffa, but what standard practice said should help people instead could torture and kill them, especially famished children.

Chris, still only twelve years old, was the only Murray not credited in this first paper. But that would soon change. Every summer between 1975 and 1980, he went back with his parents to Africa. In the Ogaden region of eastern Ethiopia, John, Anne, Megan, and Chris ran mobile clinics for sixteen thousand Somali refugees. In Comoros, a group of tropical islands in the Indian Ocean north of Madagascar, they operated schoolhouse clinics and evaluated the country's capacity for improved health care. In Kenya's Rift Valley, the family served and studied East Africa's famous seminomadic warrior tribe, the Maasai. From one very different place to another, the Murrays found new evidence that contradicted standard advice on treating malnutrition. By 1980, they had published more than a dozen papers together on diet, famine, refeeding, and disease, not only in *The Lancet*, but also the *British Medical Journal*, the *British Journal of Nutrition*, the *American Journal of Clinical Nutrition*, and *Perspectives in Biology and Medicine*. Chris's first official publication, in the June 12, 1976, issue of *The Lancet*, appeared when he was thirteen.

In Kenya, their longest and most stable location, the Murrays lived in the bush. Megan, like Nigel, was by now in college. Working directly with his parents, Chris wrote up patient histories, dispensed prescriptions, and administered basic care. As his siblings had before him, he held a flashlight for his father and learned to pass scalpel, forceps, and bandages. In the broad savannah, the teenager grew a wispy mustache and learned to drive by taking the wheel of the Land Rovers.

At home or abroad, he surged physically and intellectually. Initially quite small for his age, "a quiet young fellow" in his father's words, Chris now stood tall. His stance and gaze were confident. In his Minneapolis-area high school, where he would be valedictorian, Chris skied, ran track, and joined the debate team, delighting in demolishing opponents' flimsy reasoning. Tolkien, with his tales of small groups armed only with undaunted courage overcoming vast and powerful menaces, became his favorite author. He would become a scientist, he decided. He would study life and death and help people heal. "He became addicted to working hard," John says. "He never had to be pushed."

In Africa, at night, over dinner, sharing the simple pleasures of shade and cool water, Anne comforted Chris after difficult days. John asked what larger patterns he had observed in their patients. The sheer quantity and variety of medical problems made rigorous examination all the more important, he taught. As the entire family would write in a group letter responding to a pair of nutritionist critics, "Armchair logic plays little place in the analysis of biological phenomena; there are numerous examples of the most impelling logic in medicine which have misled physicians into perpetuating useless ideas and remedies for decades."

A hospital without doctors. Marmalade for medicine. A cure worse than the disease. *Conventional wisdom can kill*, Chris Murray concluded before his eighteenth birthday. *Science can save lives*. If our knowledge of human health was a map, it was full of false turns, missing information, and splits in the road that seemed to lead only to dead ends. To help everyone, you had to correct what was wrong and fill in what wasn't there. Like his mother, he would not be deterred from a course he thought was right, whatever its perils. Like his father, he believed the power of analysis could reveal the world as it was, not as others said it should be. He didn't know if he would follow his parents in their work, but he was already searching for his path.

CHAPTER TWO

The Third World and the Nerd World

The Save the World club—"What is the evolutionary
purpose?"—Another puzzle—Secrets and handshakes.

In 1980, Chris Murray went to Harvard for college. He was one
of those highly accomplished, somewhat eccentric high school
graduates who often end up in the Ivy League. His roommate,
Thomas Henry Rassam Culhane, was another. An Irish-Iraqi
American, raised in Chicago and New York, Culhane had left
school in eighth grade to attend Clown College. A year later, at
age fourteen, he became the youngest salaried clown in the history of Ringling Bros. and Barnum & Bailey Circus. His clown
name was Tee Hee, Attorney at Laugh.

Culhane met Murray in the Harvard freshman outdoor program, in which groups of about twelve new students go on weeklong camping trips before school begins. The pair bonded when
Murray sided with Culhane against those who told him not to
carry a guitar on their hike to Avery Peak on Bigelow Mountain
in Maine. "If he thinks he can do it, he should be allowed to do
it," Murray said. "It's his challenge." Culhane had never met anyone who was so appealingly intimidating, or as calm about his

competitiveness. With one breath, Murray said, matter-of-fact, "I'm going to get there quicker than you." With the next breath, he said, "But I think you can do it, too." Culhane, stubborn himself, with his own strong sense of identity, took both kinds of comments as a compliment. "Chris was everything I'd dreamed of what somebody could be at that age," he remembers. "He had an authority about him that nobody questioned. He'd been so many places and done so many things."

The two became best friends, and, once at school, formed a three-man "Save the World" club with another student, ardently discussing the conditions necessary to bring food, shelter, clean water, and energy to people in need. And not only discussing. First semester, Murray suggested that the threesome learn to use lathes and drills, developing skills that would be useful in the field. Culhane was incredulous. "I've come to Harvard, and I'm going to take a class in machine shop?" he asked. Yes, Murray said. "If we're going to save the world, we can't just develop our minds intellectually," he told his friend. "We have to be able to work with our hands. We have to be able to build real things."

Everything Murray did had a purpose. In line the first night at the cafeteria, he asked, "Are you going to get coffee?" Culhane shrugged. He hadn't drunk coffee before. "If you do, get it black," Murray said. "Coffee is a stimulant—a drug, not a beverage. Use it for what it's for: to stay up late." When he observed how much Culhane washed his hands, Murray told him he had to throw away his antibacterial soap. "In Africa we learned that if you bathe too much, you get sick from being too clean," he said. "You don't want kill off the good bacteria and create resistance."

There were people who didn't quite get Murray. Even in Harvard's carefully assembled collection of oddballs and outliers, he was considered strange. Murray was too intense, too confident, too indifferent to what other people thought. They kept asking Culhane, "How are you friends with Chris?" But those who shared his various passions welcomed his intensity. With Culhane, Murray was an explorer, engineer, and scientist. With jocks, he was a ski team member and ferocious intramural squash, racquetball, and rugby player. With international students, he was a person who knew the world. "He was a lot of fun to be around," Culhane says. "It was fascinating to see his social network and to be part of it."

Virtually all Harvard freshmen live on the central campus, in a large quadrangle known as Harvard Yard. For the rest of their time in school, they apply to one of a dozen "houses"—elaborate dorms, each with its own dining hall, library, social spaces, and culture. Culhane, who had joined theater and musical groups, had his heart set on one of the "art" houses, Lowell or Adams, close to both the center of campus and the Charles River. When it came time to submit their applications, however, Murray excitedly informed him, "I got us into Currier House," one of three less venerable residences built originally for female students of Radcliffe College, and located more than a mile away from other upperclassmen.

Culhane was crushed. "That's the place they say the third world meets the nerd world," he said.

"Yes," Murray said. "Isn't that perfect? You can ride your unicycle to class."

One night someone reported to administrators that Murray had shattered a huge plate glass window at the Currier House entrance. The charge was false, but it looked bad for Murray when he was found with a blowpipe built out of PVC, and nails with paper cones taped to them as arrows.

"What do you think about his behavior?" the dean of students asked Culhane in an interview. "Should he stay in school? Is he psychotic?"

The clown had to hold back laughter. "Chris is inspiring us all," Culhane said. "You don't throw somebody out of school for taking their lessons so seriously they make them real." Since Murray wanted to be able to live with hunter-gatherers as well as with other Harvard students, Culhane explained, "of course he's going to make a blowpipe," adding, "and"—because he's Chris Murray—"it's going to work."

The enormous problem Murray was already trying to solve as an undergraduate—meeting and overcoming the many obstacles to good health faced by people around the world—had many possible approaches. His parents were scientists and medical professionals. He would be, too, he assumed. But how to make the biggest possible difference? How to reach more than just the individuals one could help as a physician?

First, biology obsessed him: he and Culhane decorated their dorm room with pictures of rainforest, desert, and savannah, and shared a single hero, the Harvard evolutionary biologist and Pulitzer Prize winner E. O. Wilson, who argued that the social behavior of all animals—including humans—is shaped by genetics as much as or more than by their culture or environment. They also both found part-time work assisting Wilson Bishai, a legendary professor of Arabic. "I got a job with him entering the Koran into an Apple II Plus computer," Murray informed Culhane one day. "He needs someone else to type the dictionary." That this meant learning to touch-type in a completely new alphabet was no problem, Murray said: "You just have to rewire your brain."

Murray planned a junior year abroad for them in the Middle East. Before they left New England, however, he said, "We're going skiing."

"What is the point?" said Culhane. They needed money, and

now, just before leaving, Murray was making them rent a cabin in the New Hampshire wilderness with three wealthy ski team buddies. They lived in a penthouse, Culhane pointed out. They always treated him like a peasant. Anyway, skiing was stupid. "You go up and down," Culhane told his friend. "What is the evolutionary purpose?"

It was the question they asked of everything. And Murray was ready to answer. "When you're on the ski slope, you look down and you can decide what kind of organism you're going to be," he said. "But then when you make a choice to take a certain trail, you can't go back uphill. You've committed. By the time you reach the bottom, there's only one outcome."

"I was intrigued by this way of looking at skiing," Culhane remembers. Soon he was bundled up, poles in hand, long narrow boards strapped to his feet, riding a chairlift up a mountain. Then, in characteristic fashion, Murray, at the top, said, "See you later. See how things develop," and shot off. "So I had to learn how to ski on my own." Culhane laughs. "By the end of it, I knew how."

Murray, meanwhile, had talked up their trip, attracting potential donors by appealing to their shared sense of adventure. "You need money, don't you?" the ski team trio asked them, driving back. Each agreed to pitch in fifty dollars.

Murray and Culhane flew one-way to Paris, took a train third-class to Marseilles, and then boarded as fourth-class passengers an old steamer to Tunisia, sleeping on deck in the stormy Mediterranean. To meet their daily expenses once they arrived, they had gotten hired to research and write for *Let's Go*, the budget guidebook series based at Harvard. They went scuba diving and horseback riding, bathed in the ocean and lived on almonds washed in the salt water, in the same way, E. O. Wilson said, snow monkeys cleaned their rice. Early on, Murray found them cheap lodging in a student dorm five miles from Tunis. Getting back for dinner was a forty-five-minute jog. Feral cats roamed the cafeteria, jumping up and stealing food scraps. "It was a shocking experience for me, but Chris took it in stride, so I took it in stride," Culhane says, "living with the poorest students in the world."

One night in the dorm, as Culhane played guitar and led the group in songs, Murray met an attractive Frenchwoman their age. Agnes was five feet three, with dark hair. The daughter of an art historian and a homemaker from Clermont-Ferrand, an ancient city almost exactly in the middle of France, she was attending the same summer language program they were. While Culhane played on, he noticed the couple talking. They wandered off on their own, away from the others, and soon were spending more and more time together.

The summer course ended. Agnes returned to France, promising to keep in touch. Culhane went to Cairo, where he taught English, played in a rock band, and trained with the Egyptian circus. Murray explored other parts of Egypt for *Let's Go*, and then moved on to Pakistan and India. A year passed before the roommates reunited in the fall of 1983, their senior year at Harvard. Their experiences had changed them, Culhane would remember. "We'd understood the poor by living among them," he says. "We weren't just kids spouting off."

Back on campus, E. O. Wilson—soon "Ed" to them, though he was three decades older and one of the most eminent scientists in the world—became Murray's mentor and would advise him on his senior thesis, which analyzed the number of species a given area could support as a game or nature reserve. Wilson, in successfully recommending Murray's work for Harvard's highest undergraduate thesis prize, called it "brilliant and with many potential applications," based on "highly original research that could constitute a substantial part of a Ph.D. thesis." But Murray was already moving from biology to new interests. Harvard students, for the first time, had personal computers in their dorm rooms. Murray commandeered Culhane's. "The future is, you're going to be able visualize *everything* on a computer," he said. Economics equally entranced him. They *had* to take courses in the econ department to know how the world is run.

"I don't like how the world is run," Culhane said. "If you take these classes, it will change you."

"Not me," Murray said. "I can take these classes and find new ways of seeing and interpreting things."

He started bookmarking economics textbooks for Culhane, urging, "You've got to read this."

Culhane was completely put off. "It's just numbers and graphs," he told Murray.

"It's badly done," Murray conceded. To get the big picture, you had to imagine things the numbers and illustrations only suggested. He reminded Culhane about the people they had met in the Middle East, and told stories of assisting his parents, the bush doctors, in villages across Africa. From a single graph about poverty, he traced children suffering, parents struggling, families trying together to improve their health and fortune. "He's like a seer," Culhane thought. "He can see the past and the future." The numbers came to life for Murray. "For me," says Culhane, "they stayed on the page."

His friend, Culhane thought, was finding the vantage point from which he could survey continents. "Chris was never competing with anyone else on campus and certainly not with me," he says. "It was always about improving himself relative to his family." Murray spoke with such admiration about what his parents and siblings did. He wanted to prove he could be as dedicated and effective in improving the world.

In January of his senior year, Murray was selected as a Rhodes scholar, one of the most prestigious awards there is, funding study at Oxford University. After one last summer in Kenya with his parents, he arrived in England in the fall of 1984.

As an academic setting, Oxford was in some ways Harvard's opposite. Here one's social life was very structured and one's scholastic life very unstructured. For Murray, it was a perfect fit. He skied, played squash and cricket, dined at High Table, and hung out with other fellows at Rhodes House, and befriended, among other students, the Crown Prince of Japan. Agnes, the Frenchwoman he had met in Tunis, got into an Oxford graduate program in languages. His favorite author, J. R. R. Tolkien, had been a professor at Murray's college at Oxford, Merton, in the 1940s and '50s, and his presence was still palpable. Murray roamed "the sundial lawn" and other quaintly named settings, ideal for rumination, deciding what to do with this opportunity. In theory, he was at Merton studying international health economics. In fact, he was charting his future life.

Say you wanted to make the whole world healthier, Murray asked early in his research for his Oxford doctoral dissertation. How would you do that?

A decade earlier, the answer had been simple—at least to economists. Sick people tended to be poor. To get better, they just had to get richer. Poor people were hungry—money bought food. Poor people washed and drank near where they defecated—money bought plumbing. Poor people lacked medical care—money bought vaccines and the trained men and women in bright white coats to deliver them. The doctor and demographic historian Thomas McKeown, who had studied declining mortality in England and Wales from 1850 to 1970, typified the current thinking. Compare developed and developing countries, said McKeown, and "There is little doubt that the differences in health experience are attributable mainly to the direct or indirect effects of poverty, and would be largely eliminated if it were possible to raise the lower standards of living and medical care to the level of the highest."

Then opinion changed. A few intrepid researchers ventured outside the library, laboratory, and classroom. They visited primary health care projects around the world. There they discovered something both completely obvious and largely ignored. *Not all poor people were the same.* Yes, all had relatively little money. But reports repeatedly heralded residents of some low-income countries—China, Costa Rica, Sri Lanka—and the state of Kerala in India as healthier than others and, in terms of mortality rate improvement, doing even better than those in many wealthy Western nations. According to the World Bank, for example, China, Sri Lanka, and Kerala each had per capita incomes of, at most, $330 in the early 1980s. Nonetheless, in all of them life expectancy at birth approached seventy years. In Costa Rica, income per capita was $1,020, an order of magnitude less than in the United States. Yet the two countries' infant and adult mortality rates were about equal.

The old model of health improvement based on economic growth had been dubbed the "Northern Paradigm." The new model, called the "Southern Paradigm," was centered on equal access to health care, education, and nutrition. For instance, the Rockefeller Foundation would publish a highly influential 1985 report, "Good Health at Low Cost," promoting China's patriotic health campaign, Costa Rica's universal health insurance, Sri Lanka's land reform movement, and Kerala's rural nurse-midwives as reasons residents of these areas were living better. Hard data was sketchy, however, and largely limited to average life expectancy. Who was really extraordinary, Murray asked, and what specific interventions did the greatest good?

At the library, he pulled every statistical compendium he could find from UN agencies and the World Bank, the two leading sources of international health data. He wanted to understand two things. First: How do you come up with a comprehensive summary of the health of a population, so you can actually say, "Sweden is healthier than Canada," or "Niger is less healthy than Nigeria"? Second: What specific factors lead some countries to do better than others at the same income level? "How do we know who's really exceptional so we can replicate that experience?" Murray wondered. "What is the strength of the evidence, piecing together data in a really imperfect world?"

He already knew the limits of individual on-the-ground interventions. Now he saw enormous gaps in the information prepared by agencies at the highest level. By far the most popular single index of health status for health planners and economists was a country's infant mortality rate, for example. As health initiatives went, helping children survive past their first birthday was a very good one. But you didn't want to mistake progress in one area for overall gain. Murray graphed the most trustworthy national life expectancy figures against the corresponding infant mortality rates for the same countries. The graphs suggested only a vague association, not the tight line of correlation. A child named Betsy might be born in Uganda and a child named Bill might be born in Ethiopia. Even if both countries had the exact same infant mortality rates—and both children survived to young adulthood— their life expectancies might differ by more than ten years. Did Betsy and Bill have the same health care needs?

Other data didn't make sense in all sorts of ways. Only a handful of countries in the developing world had complete vital registration systems—that is, they had birth, death, and census records for at least 90 percent of their population. How were basic health statistics being generated for everyone else? Murray counted at least five separate models for estimating life expectancy used by UN agencies and four by independent demographers. Their results varied by as much as fifteen years. The Population, Health, and Nutrition Department of the World Bank said life expectancy in the Congo between 1980 and 1985 was 60.5 years, for example. The Estimates and Projections Sub-Division of the UN Population Division said it was 44 years. For Namibia, the difference was 12.2 years; for South Africa, it was a decade.

Depending on which estimate was used, one expert's basket case might be another's star performer. Between 1981 and 1985, for instance, Bangladesh, Bhutan, Burma, the Congo, Mongolia, and North Korea all made academic expert lists of "exceptional" or "best" performers in improving longevity. Attention and funding followed for public health programs based on their examples. Should leaders focus limited time and money on safer birth conditions or lower hospital copays? On improved nutrition or better drinking water? What if none of the above was as important as the number of years young women spent in school? Policy makers were making decisions and allocating resources based on fragmentary and contradictory information, much of which was at best a guess.

Work in international public health might be the best way to make our planet healthier, happier, more prosperous, and more peaceful, Murray thought. It was arguably the most important work in the world. But how could we say what would make people live better if we couldn't even be certain when and how they were dying?

Oxford had held classes since 1096. Harvard was founded in 1636. The leading international institutions from which Murray was getting his data were much younger, each formed in the aftermath of World War II: the World Bank, based in Washington, D.C., in 1944, to promote economic development; the United Nations, or UN, headquartered in New York City, in 1945, to broker political agreements; and the World Health Organization, or WHO, established as an agency of the UN in Geneva, Switzerland, in 1948, to coordinate, advise, and aid national health systems and public health programs. In January 1985, Murray traveled to Geneva with two other Rhodes scholars. He wanted to know if the problems he was seeing with international health data were real, and, if they were, how they could be fixed.

A former Rhodes scholar working at the WHO set up a grand tour for the visitors. "It was the most incredibly educational trip," Murray would recall. "Here we were, three kids, trotting around, meeting all the senior brass of the organization"—the director-general, assistant directors-general, and numerous division heads. If anything, these eminences were too high up in office to be able to answer the twenty-two-year-old's specific questions, or even to know who below them had produced the numbers so perplexing to Murray when he found them in the library. Still, he learned plenty.

WHO headquarters, built in the mid-1960s, is a wide, nine-story building with a glass-and-aluminum facade. From a distance, bordered by green hedges and set on a sloping lawn, it resembles a giant rectangular terrarium. Inside, public areas are airy and attractive, filled with casual seating and decorated with abstract art, busts of past public health leaders, and statues and carvings donated by member countries. Conversations in a dozen languages echo off high concrete walls and a marble floor speckled in different shades of white, black, orange, and brown. Murray's tour ended at the attached executive board building, a stone cube decorated inside in bright '60s orange. When the executive board meets, the various national ministers of health elected to guide and implement decisions of the larger WHO Health Assembly sit at a forty-seat circular table.

Upstairs in the spectators' gallery, Murray and his friends eagerly grabbed the earpieces that offered simultaneous translation. Even in English they could barely understand what the delegates were arguing about: it was in UN bureaucratic speak. Their host, though, translated the translations with a candid running commentary: "He wants funding," "She's trying to get her boss's job," "He's saying he'll switch votes if they do, too." It was all new to Murray. He realized that the broadest and best-funded initiatives to improve world health did not resemble individual efforts like those he'd participated in with his family or imagined with Culhane. They were about politics and diplomacy, promises and threats, secrets and handshakes.

Murray met different international leaders of programs fighting measles, malaria, and other major diseases, and he peppered them with impatient—almost impertinent—queries. His host, sympathetic if puzzled, took him to the deputy director of the Division of Health Statistics, Ian Carter, an Australian. Murray repeated his questions in even curter fashion. He explained that he had tried to understand where the numbers for mortality in Africa and other developing areas were coming from, but to no avail. Carter looked him up and down. A Rhodes scholar interested in death data details? As it happened, there was another crazy-smart guy in Geneva, barely ten years older than Murray, asking the same kinds of questions. "The person you need to meet is Alan Lopez," Carter said.

It would be the start of three decades of collaboration.

CHAPTER THREE

How to Die with Statistics

The weaker sex—"Are any of these the same deaths?"—An
invigorating conflict.

afeguarding the health of the world depends as much on math
as on medicine, and people who can gather and understand big
sets of numbers are essential to the workings of public, as opposed
to personal, health. Start with the science of epidemiology. The
name derives from Greek roots meaning "the study of what is
upon the people." Unlike most doctors, epidemiologists deal not
with individual patients but populations at large—what makes
them sick, how diseases spread, and how these maladies may be
controlled. Epidemiology originated as a formal discipline in the
late nineteenth century. The classic example of the successful use
of medical statistics comes from the 1854 cholera epidemic in
London's Soho neighborhood. Decades before the formulation of
the germ theory of disease, Dr. John Snow mapped the outbreak
of cholera cases, identified a cluster around the Broad Street public water pump (later shown to be installed beside an old cess-
pit), and convinced authorities to remove its handle, making the
pump inoperable and thus cutting off the key local cause of the
transmission of the disease.

To later epidemiologists we owe our knowledge of the benefits
of hand-washing and the disinfection of surgical instruments, the

links between tobacco smoking and lung cancer and between sexual intercourse and HIV/AIDS, and how SARS (severe acute respiratory syndrome) spread so rapidly during a 2002–2003 global outbreak. In each instance, physicians and biologists were responsible for identifying why and how these diseases proliferated, but it was someone using statistics who first had to show the world what was happening, where, to whom, and any outside factors—whether a shared water pump, grubby fingers, cigarettes, a sexual partner, or something else—the cases had in common.

Alan Lopez, the young WHO researcher Chris Murray was sent to see, had discovered epidemiology after he earned a bachelor of science in mathematics in his native Australia and went to Purdue University, in Lafayette, Indiana, in 1973, for a master's degree in statistics. "I was really intrigued by this," Lopez would remember. "I was interested in applications of statistics, such as econometrics. I thought maybe I should work in the banking or financial industry, but I wasn't that motivated about it. But I was passionate about the application of statistics in medicine."

He found himself shut out from Ph.D. programs in epidemiology because he was not a medical doctor. The next best thing was demography—the study of populations, beginning with birth and death rates. The demography program at the Australian National University (ANU) in Canberra, his country's premier research institution, offered a medical subfield. Close enough, he thought. Finishing his year at Purdue, Lopez traveled back to the other side of the world and ended up writing his Ph.D. thesis on 125 years of changing mortality in Australia.

At the beginning of the twentieth century, his calculations showed, men in Australia lived nearly as long as women: the gap between them was just three years. Then, thanks to increased access to family planning services, better prenatal care, and more and improved deliveries in hospitals, many fewer women died in childbirth. Meanwhile, a larger percentage of men took up heavy smoking, reduced their physical activity, and adopted unhealthy diets. By the latest point in Lopez's data set, 1975, Australia, like the United States, was facing an onslaught of male cardiovascular disease. Mortality among men was rising for the first time in recorded history without an accompanying war or epidemic. The gap in life expectancy between men and women was now seven or eight years.

Lopez's thesis—entitled "Which is the weaker sex?"—laid out the numbers, relying on epidemiology for explanations. This was daring. Demographers traditionally cared only about *rates* of mortality by age and sex, not their *causes*. What number of middle-aged men die annually, they asked—not whether they die from heart disease or homicides. And epidemiologists generally focused their efforts only on a specific outbreak, not to shifting country-and-continent-wide trends. Why are men dying of heart disease, they asked, but not how many men die overall.* Lopez refused to accept the distinctions between disciplines and definitions. What mattered to him was that people were dying. "I was in this very interesting area between demography and its description of population changes, including mortality, and epidemiology, which tries to explain mortality," he would recall. It was the what and why of death, at last forced to speak to each other.

Soon after graduation, Lopez moved to Geneva—by no coincidence the preferred location of his girlfriend, Lene, a Danish Ph.D. student he had met while at ANU—to work at the WHO's Division of Health Statistics on a global study of sex differences in mortality, the exact topic of his thesis. The initial contract was for three

---

*Again, a look at the derivation of terms is revealing: an epidemic is a plague that comes from outside—*epi*, upon—a community, strikes in force, and then subsides. This is in contrast to endemic diseases, which are native to (*en*, within) a certain area, striking the local population regularly.

months, beginning August 1980. Approximately 1,500 professionals then worked at the World Health Organization, with some 70
in the Division of Health Statistics. Yet Lopez, as a young grant-
funded researcher working on a temporary contract among WHO
lifers, was free to pursue his own approach to research. "I was not
content to limit myself to demography," he says. "It was mostly involved in the measure and description of population phenomena—
death rates, fertility, censuses—but it didn't go into explanation."
He wanted to know: *Why* was lung cancer rising? *Why* was heart
disease rising? His new boss, treating him like a visiting scientist,
gave him license to explore these kinds of questions.

Working at the WHO didn't make excessive demands, and it
conferred a number of benefits. Any expert in the world he wanted
to speak with, Lopez could now just call, and he or she would
answer his questions, or, better yet, offer to meet in person. "Go
out and create knowledge," he was told. "Take all these databases
the WHO produces and make use of them." And, unlike almost
every other member of the WHO staff, then and now, Lopez was
allowed to publish what he discovered, under his own name, in
independent journals, because everyone thought he would return
soon to academia.

Other WHO employees around him were assigned to the
same task or specialty, year in, year out. Lopez roamed unfettered.
"If I'm going to work and have an impact in health, the WHO
is the place to do it," he thought. His grant was renewed, and he
decided to stay. His work kept evolving. From the study of sex
differentials in mortality, he branched out in the early 1980s to
projecting the number of doctors and nurses globally, estimating
socioeconomic inequalities in Europe, and analyzing the health
of the elderly. "I was busy, very busy, traveling a lot, meeting and
connecting," he remembers. "That set me off on a path quite different from most of the other people at the WHO."

By 1984, Lopez had become a permanent staffer. Still, the larger bureaucracy was such that Lopez had to ask permission to write to the designated main statistical agency in each of the WHO's 190-plus member countries so that the organization, for the first time, would have a compendium of all available data on mortality and causes of death. "That project reflected my desire and curiosity not only to describe health problems in rich countries and regions like the U.S. and Europe and Australia," he says, "but to describe, as best we could, health problems in poor countries, because that's where the investment had to be made." Once many Australian women died in pregnancy; now few did. Once few Australian men died of heart disease; now many did. What was happening to women and men in Bangladesh and Indonesia, Ghana and Peru, Kazakhstan and Papua New Guinea?

Once approved, Lopez took the numbers and began prepping them for analysis. Tracking death in Australia, Lopez would later say, had been a position of relative safety. Trying to do the same thing for the entire globe—"taking very scrappy pieces of information from all around the world and trying to make sense of them"—was a statistical high-wire act. Look down long and you'd be dizzied by what you didn't know, or maddened by inconsistencies like those Murray had turned up in Oxford.

Lopez began with the fraction of the big picture deemed most important to international health, and therefore most studied: children who died before age five. Then as now, "the global public health community was very much focused on child survival," he says. Compared with the diseases of adults, the most common killers of children were much more easily prevented with a small set of inexpensive interventions (for example, growth monitoring, oral rehydration therapy, breast-feeding, and immunizations) and doing so seemed to promise a lifetime of productive participation in society. To save the most lives, though, you had to know how exactly children were at risk. Keep it simple, Lopez told himself. Find out: *Who dies of what?*

That so fundamental a question—repeat, who dies of what?—was new terrain spoke to the enduring chasm between professional epidemiologists and demographers. In the United Nations system, they were even separated by the Atlantic Ocean.

The Division of Health Statistics, where Lopez worked, was a tiny part of WHO operations in Geneva. Much bigger—90 to 95 percent of the total staff, Lopez estimated—were units devoted to the control of specific infectious diseases, the promotion of newborn and maternal health, and nutrition. One such program set policies on hand-washing, water chlorination, and oral rehydration therapy to fight diarrheal diseases worldwide, for example. Another addressed malaria through insecticide spraying, the distribution of protective bed nets, and large-scale draining projects to remove the habitats of parasite-transmitting mosquitos. A third promoted immunizations against measles and tetanus. A fourth combated childhood pneumonia via immunization, vitamin supplementation, clean air and water campaigns, and education programs to promote breast-feeding. Noncommunicable diseases—cancers, heart disease, and chronic lung conditions—were the domain of a small separate group. Smallest of all—just one person—was the staff on injuries. And each group of epidemiologists produced independent estimates of the annual number of child deaths attributable to its specialty.

Meanwhile, the United Nations Population Division in New York, made up of demographers like Lopez, trained in statistics, took all available data—surveys, censuses, health reports, and government figures—and produced separate estimates of the number of child deaths in the developed and developing world.

In short, one camp, spread across different disease control

programs in Geneva, said how many children died by cause, while the other, based in New York, said how many children died in rich and poor countries. Seemingly no one before Lopez had thought to compare notes and see if the two ways of looking at life and death could be combined in a common, comprehensive view of the world. The initial math, though, couldn't be simpler. On one piece of paper, Lopez added up the different disease-specific child mortality estimates of the WHO. On another, he wrote down the UN's estimate of total child deaths in 1980. The first number was close to 30 million. The second was under 20 million.

Uh-oh. "When I put them next to each other, I was getting substantially more deaths adding up the WHO figures than those provided by the UN," Lopez would recall. Fifty percent more, to be exact—and as much as *200* percent more in certain areas of Africa. The leading global authorities on human health were either missing or inventing 10 million dying children a year.

Both WHO and UN estimates had their problems, but, if the choice were between epidemiologists or demographers, Lopez had no doubt that the demographers' figures, the ones from New York, were more credible. "There was a lot more data collected on the *fact* of death rather than the *cause* of death, which requires a medical doctor to certify what the child died from, and medical doctors were in short supply," Lopez says. And epidemiologists working separately could easily fall into the error of double counting. Children who were malnourished, for example, also had a higher risk of having pneumonia *and* diarrhea *and* measles *and* malaria. The total number of child deaths estimated by the demographers at the UN had to be closer to the truth than the sum of child deaths by different causes put forward by the separate WHO staffs.

"These were well-meaning advocates doing the calculations, but because there was no central oversight at the WHO, different groups were using different methods, different degrees of rigor, and databases of different quality," Lopez remembers. "When they made their estimates, there was no constraining them. No one was saying, 'Hold on a minute, you're saying five million deaths from diarrhea. The people down the corridor are saying five million deaths from pneumonia. Are any of these the same deaths?'"

Such double counting could have fatal consequences of its own. The whole purpose of making estimates was to correctly describe the comparative importance of different diseases, particularly in places where medical professionals and supplies were extremely limited. Diarrhea and pneumonia were the leading causes of child deaths in the 1980s. They had two completely different treatments. "So if you had nine million deaths between the two of them and you said there were four or five million each, but in reality there were three million from diarrhea and six million from pneumonia, you would end up with a lack of pneumonia treatment," explains Lopez. "That potentially could cause many more child deaths."

Add up deaths attributed to just four causes—diarrhea, pneumonia, malaria, and measles—and you had more deaths than the number of dead children. "It didn't even include that children were dying from congenital anomalies," Lopez remembers. Or cancers. Or birth trauma, malnutrition, fires, falls, drowning, or car crashes. "Even without including any of these we were exceeding the total."

Lopez brought his concerns about inaccurate and double counting to colleagues. The head of the diarrheal disease program, a fellow Australian, expressed interest in Lopez's findings. Everyone else, though, essentially disregarded him and continued to promote his or her own statistics. "I remember shaking my head and thinking, 'This is good evidence and yet these programs do not want to engage with me," Lopez recalls.

For the first time, he realized the limits of his status as a researcher in the deep recesses of the WHO bureaucracy—an outsider on the inside. Even after he went back to the data, meticulously combining small studies in different parts of the world to avoid double counting, the reaction was, basically, *Alan, mind your own business*. "I was a young scientist," Lopez says. His boss still supported him, but couldn't overrule the disease control program heads. "They thought that I was a menace and wished I would just go away."

In academia, the rule is "Publish or perish." In a bureaucracy, it is "Don't embarrass the higher-ups." As much as it annoyed Lopez to have his scientific logic ignored, he understood that for WHO program purposes—for raising money, for raising awareness of the tragedy of child deaths—overestimating could be helpful. The more children who were said to be dying, the logic went, the more likely donors and the general public would care. And even if overestimating one disease inevitably meant shortchanging treatment of another, decision makers might start to lose faith in the WHO if their numbers suddenly changed. "If one year you said there were three million deaths due to pneumonia instead of five," Lopez observed, "people, including the donors, would say, 'What are you doing?'" It was wrong to knowingly inflate numbers. But at stake either way were the lives of millions of very poor children.

Lopez was considering whether to risk his career by publishing corrected estimates without institutional support when a young man knocked on his door.

"Is this Alan Lopez?" the stranger inquired.

"I said yes," Lopez remembers. "He said, 'My name is Chris Murray, and everything you've written about mortality in Africa is wrong.'"

Alan Lopez was an affable man, politic and polite. Chris Murray was neither. But Lopez took an immediate liking to Murray, the younger man's brashness notwithstanding. Their shared roots in the Antipodes—the simultaneously pretentious and condescending British term for Australia and New Zealand that means, literally, the opposite side of the earth—made for a natural rapport. They understood each other. Most important, they shared a common obsession: finding out what was actually killing people around the world, and why. It felt good to Lopez, finally, to have someone pushing him forward rather than holding him back.

The two compared findings and began corresponding. "Chris was at that stage agitated by all these differences in international mortality statistics," Lopez would remember. "He didn't think this was reasonable. He was quite right."

Back at Oxford, Murray widened his dissertation research to cast a critical eye on the statistics of the WHO, the UN, and the World Bank. Traveling to each organization in turn, he was given free access to their findings, but learning their sources and methods was another matter. "I wanted to know: how did they know X about country Y, or Y about country Z?" Murray remembers. "I got exposed to the UN evil bureaucrats in a major way. I was stonewalled. They wouldn't tell me anything."

In the end, a little mathematical reverse-engineering revealed what his human sources, at least at first, would not. In instances where UN and World Bank estimators had no new information, Murray discovered, they just assumed a steady life expectancy increase of two, two and a half, or three years improvement every five years, up to a life expectancy of 62.5. Someone had come up with the formula in 1955, and thirty years later this was still how most estimates of life expectancy at birth were being made. Murray counted thirty-eight countries in Africa whose UN estimates since 1970 followed this same model exactly. (Where UN and World Bank estimates differed dramatically in Africa, it was because the World Bank wove in separate—but not necessarily more reliable—figures from a third organization, the United Nations Children's Fund, or UNICEF.) This wasn't serious statistical analysis based on real experience. "According to the UN," Murray wrote in a scathing paper summarizing his work, "no country in the world has or will experience a fall in life expectancy between five year intervals at any time between 1950 and 2025, an assumption known to be untrue."

The UN Secretariat actually published multiple sets of life-expectancy estimates. Of these, the most used and most quoted source was the UN *Demographic Yearbook*, which reported official government figures from UN member states as submitted to the UN Statistics Division. The *Yearbook* was a standard reference in libraries and the common starting point for all variety of researchers. But governments without vital registration systems—or professional statisticians—had even less ability to make accurate estimates than the UN and World Bank. Many collected almost no information on birth and death; others didn't like their data and made it up. Botswana, for example, might report its infant mortality rate was the same as Italy's. "The numbers in [the *Yearbook*] are not evaluated for their likely validity or even internal consistency," Murray found. "They are simply published." It was like bartenders asking drinkers to report their own levels of inebriation before last call: some people will be honest, others will lie, and most will have no idea either way. Yet only if a country didn't return the questionnaire was a more scientific estimate printed.

What was the result of doing things this way? In 1982, the *Demographic Yearbook* reported that life expectancy at birth in Pakistan was 51.8 years, as estimated by the UN Population Division. A year later, though, the number was 59.1 years—an almost-decade leap—because that was the figure now submitted by Pakistani officials. In 1985, the *Demographic Yearbook* said that life expectancy in the Gambia was 43 years, as estimated by the Gambian government; the next two years, the UN Statistics Division received no response from the same government, and so published the Population Division's estimate of 33.5 years—an almost-decade *drop*. It was laughable, but a *Yearbook* technical note said, since the numbers provided came either from government officials or from estimates prepared at the UN, "they are all considered reliable." It was enough to make Murray weep.

But instead he raged. "Estimates are done by individuals who impart their own style and beliefs to each estimate," he would write. "There is no uniformity in technique, from country to country, or assessment to assessment. Because no empirical data exists at all for a number of countries ad hoc techniques have been employed." The most-trusted authorities in the field used made-up math or ridiculous but "official" government statistics. Yet life-and-death policy decisions—whom to save and how— were based on these demonstrably false sources.

Murray's paper exposed the flimsy, almost arbitrary basis of these respected statistics—and therefore any public health program that depended on them. Without accurate basic information, the world could not identify—much less copy—leading countries' best practices. Health officials from Iceland to India had no idea how to direct needed resources to where people were sick and dying in the greatest number. At the highest levels of world health, people were constricted by politics, divided in their efforts, and beholden to national interests that kept them from saving lives.

Lopez, given a draft of the paper to read, thought it so incendiary he was surprised the pages didn't self-destruct. "Chris was arguing quite coherently why those estimates were inconsistent,"

Lopez remembers. "But I spent quite a bit of time trying to get him to depersonalize it—to take away the names of the people who were involved and just talk about institutions. I felt he was a very bright young man and he had a huge career in global health, but that he might shoot himself in the foot on day one."

What was unusual about Murray—inspiringly or infuriatingly so—was that he found both collaboration and conflict invigorating. Starting a debate with high-ranking officials didn't bother him, and neither did bucking established protocols. The experience in Diffa, when he was ten years old, had been his "Emperor's New Clothes" moment: what others called aid was really an empty hospital; what they called medicine was rotten marmalade; what they called a healthy diet was also nourishing malaria. The best way to save lives was using scientific methods to test accepted truths, he believed, even if it meant taking on the leading institutions in international health. "I had various problems with how assumptions were being made," Murray says. "Most people just thought these were the truth."

At last, he yielded to Lopez's counsel: the names went from his paper. But only the names. "The ready availability of data on life expectancy, infant mortality, and child mortality does not ensure its quality," Murray wrote. "Rather, it is further proof of the widespread demand that exists for such information." In 1987, *Social Science & Medicine*, a highly reputable journal, accepted the article, his first publication since the last Murray, Murray, Murray, and Murray productions of 1980. By the end of his time at Oxford, his sense of self and belief in his life mission—to measure how we sicken and die in order to improve how we live—were absolute.

The "movement of transparency," as Lopez termed it, had begun.

CHAPTER FOUR

Missing Persons

10/90—A staggering example of neglect—"You could
actually see things change"—No-man's-land—Treating the
entire world.

Chris Murray graduated from Oxford in 1987 with a doctorate
in international health economics. Just after, he and Agnes, his
French girlfriend, got married in Clermont-Ferrand Cathedral, a
grand Gothic edifice built with the region's large black volcanic
rocks. For the sake of Agnes's family, one of the officiants was a
local Catholic priest. The other was the minister of Westminster
Presbyterian in Minneapolis, the church that had helped sponsor
the Murray family's 1973 trip to Africa. Chris and Agnes then
moved together to Cambridge, Massachusetts, where Chris was a
first-year student in Harvard Medical School.

All three Murray children who had gone with their parents to
Africa would become medical doctors. They were all also multi-
taskers on a global scale, and forever shaped by their childhood
experiences, though each in a different field. From Dartmouth,
where he'd majored in geology, Nigel had gone to medical school
at the University of Otago in New Zealand, John and Anne's
alma mater. In exchange for his subsequent service, the New Zealand Army funded his master's degree in occupational health at
Harvard. By the late 1980s, he was completing his residency at

the Royal College of Physicians in London, working out of a military base attached to a medical research unit. Soon he would be sent by the New Zealand government to help provide health care in Iraq and Bosnia. "I could see my parents in the midst of all these dying and thirsty people, with lots of disease, and they got to work," he remembers. "They opened the truck and said, 'Okay, go.' You could say, 'Oh, this is shocking.' You could retreat into yourself. In getting to work, you got through that."

Megan majored in philosophy in college, also at Dartmouth, but got a job after graduation administering a refugee camp in Thailand. She was there for four years, but craved the closer personal connection to patients her father had had as a physician, and she entered Harvard Medical School the year before Chris, in 1986. In Asia, as in Africa, one of the biggest barriers to resettlement for refugees was the communicable diseases they carried, and she made studying tuberculosis and its treatment her specialty. "We were there as witnesses to just extraordinary neediness," Megan would say of her family's work in Africa. "It was obvious that we weren't making the impact that we might have been if we had had the right equipment."

When Chris arrived, Harvard had just switched to a new case-based method for medical education, now standard nationwide. Fewer hours in a lecture hall—a "mere" four a day—meant "more time on your own to learn things," he was told. Murray translated this to mean: "I could keep going on what I'd started with my doctorate." Soon, doing just that, he had talked himself into a job with the Harvard Center for Population and Development Studies, commonly called the Pop Center. Pop Center faculty included demographers and epidemiologists, economists and philosophers, physicians, engineers, environmental scientists, anthropologists, visiting foundation executives, and leaders of international non-profit organizations. In weekly seminars, one expert or another presented his or her latest work to all the others. Murray found the overall environment "incredibly dynamic." "The fact that I could hang out there during med school was fantastic," he would remember later.

His office was not in the Pop Center proper, a renovated three-story Victorian house on a red brick side street off Harvard Square. Rather, the new research fellow walked next door through a broken triangle of sloping asphalt to a cramped, badly lit one-story World War II–era cinder-block annex dubbed "the Bunker." Within a few days of starting on the job, he could be found there almost any time he wasn't at medical school, day or night. While Agnes set up the couple's new apartment and tried to learn her way around her new country (she eventually found work at Harvard's Peabody Museum of Archaeology and Ethnology), Chris was poring over documents to the music of the low ceiling's ever-clanking pipes.

The Pop Center hosted an independent international initiative, the Commission on Health Research for Development, the brainchild of more than a dozen philanthropic leaders. Lincoln Chen, then the Pop Center director, told Murray that research was undervalued. "Global health needed a big boost," Chen recalls of that period. "Most people thought research was just people running around with white coats in a lab with rats. We defined it in a much broader way, including a mother testing different cough syrups on a child, or a farmer testing different seeds and seeing what grows better."

Murray's contribution was tracking the health problems scientists were trying to solve and then comparing those to the health problems people worldwide actually had. Ailment by ailment, he broke down research funding from foundations, the U.S. National Institutes of Health, European governments, and the Japanese government. It was an approach no one else was taking. Apart from Alan Lopez, demographers didn't look at global causes of death. Epidemiologists didn't look at financial contributions. Donors didn't look at the precise etiology of disease. And medical students, as a rule, didn't take on enormously time-consuming second jobs that had little to do with their classwork. Murray, an economist-turned-physician, equally well versed in statistics and social policy, was one of a kind.

In his windowless office, the young man jotted quick calculations, including one he showed Lincoln Chen one afternoon in the Bunker hallway. The "10/90 gap," Murray called it. People in developing countries endured more than 90 percent of the world's health problems. Those problems, however, received less than 10 percent of health-related research investments. If you had money and diabetes mellitus, the drug development system was working to find a cure for you. If you lived on less than a dollar a day and had a hookworm infection, you would suffer just as much if not more; it was just unlikely that any effective treatment would be available to you.

Over the course of two years, members of the Commission on Health Research for Development would meet eight times in as many countries. They would solicit testimony and advice from hundreds of local and international experts. They would produce ten case studies. Yet Murray's early back-of-the-envelope discovery came to define the entire project. The more precise results were even worse than his tentative one. As the commission's report, released at the 1990 Nobel Symposium in Stockholm, put it: "Our most striking finding is the stark contrast between the global distribution of sickness and death, and the allocation of health research funding. An estimated 93 percent of the world's burden of preventable mortality . . . occurs in the developing world. Yet, of the $30 billion global investment in health research in 1986, only 5 percent or $1.6 billion was devoted specifically to health problems of developing countries." What Murray had first spotted, using his new numbers, became a headline story in capitals around the globe. International aid couldn't just be the distribution of existing cures. It needed broad innovation. The gap inspired task forces and global forums for health research, hundreds of papers, and international conferences by the dozen.

Eminent scientists tend to have huge egos. They need enormous self-confidence just to make themselves heard over the noisy chatter of global scientific exchange, particularly if what they are saying demands a major shift in conventional thinking. Murray's ego was as healthy as anyone's; in this case, though, he had let his data do the talking. "Chris was very steady, very methodical," recalls Lincoln Chen. "He didn't declare that he had a breakthrough. He didn't personalize ever. He allowed the evidence to speak for itself."

"It was a meaningful result," Murray would remember. "It showed just how skewed the allocations of resources were." The economist in him wasn't surprised. People were selfish. They spent their money on themselves. Yet saying so out loud, not as a moral accusation, but in stark unbiased numbers, he learned, could change things.

"Nobody knew if anyone would pay any attention," he said later. "It turned out they did."

One offshoot of the Commission on Health Research for Development report was the identification of specific diseases whose cures demanded new investment. With tuberculosis (TB), the same disease to which his sister Megan would devote much of her scientific career, Murray discovered a staggering example of neglect.

TB is caused by the bacterium *Mycobacterium tuberculosis*.

Nodular lesions called tubercles grow in the lungs, bones and joints, and central nervous system. As organ and tissue cells die, the bacteria spread to a density of as many as 10 billion per milliliter. Patients grow weak and feverish. Their chests ache. They cough, and, in the worst cases, spit up blood as they slowly die.

In wealthy countries, new treatments and intensive control measures had largely tamed this disease that in the nineteenth century killed as many as one in ten people. In developing countries, on the other hand, the problem persisted as a fact of life. In the late twentieth century, tuberculosis, spread by a simple cough or saliva, infected somewhere on the order of 7.1 million people annually: approximately 5.4 million North Africans and Asians, 1.2 million sub-Saharan Africans, and 540,000 South Americans, Central Americans, and Caribbeans, Murray estimated. Every year, more than 2.5 million of those men and women died.

No other single pathogen killed as many people, he concluded. And the number of cases would rise rapidly, since those most vulnerable to TB were the growing numbers of people also infected with HIV—not children, mainly, but adults. "These are the parents, workers and leaders of society," he wrote. All the same, international health researchers ignored tuberculosis almost completely. A 1986 Institute of Medicine study, for example, had classified diseases into three levels of priority for vaccine research. Leprosy, which was much less common, Murray noted, "received significant attention." Tuberculosis "was not even mentioned in the lowest priority group." There was just one person working on TB in all of the WHO.

What made this even more maddening was that early interventions against the disease were both effective and cheap. In select countries like Malawi and Tanzania, Karel Styblo, the world's foremost tuberculosis control specialist, had reported cure rates approaching 90 percent using existing diagnostic technology and short-course chemotherapy. Treating the most common cases cost less than $250 per death averted. Put in terms of cost per year of life saved, the tab wasn't even $10. Murray teamed with Styblo and another collaborator from the International Union Against Tuberculosis and Lung Disease to publish this finding in the *Bulletin* of that organization in March 1990. "We estimate that the total increased cost of treating all new cases of tuberculosis through a well-managed chemotherapy program to be less than 700 million US dollars per year," they wrote. By comparison, Murray later calculated, *not* treating TB would cost Americans alone $4.1 billion before the end of the decade. He strongly recommended Styblo's short-course chemotherapy treatments be applied worldwide.

Why didn't anyone say so earlier? "At the time, TB was really complicated and confusing," Murray remembers. "There was this old literature. People working on it had their own language. They only spoke to each other." Medical research—and funding—got siloed. Publicity was falsely equated with importance.

Murray, though, spoke many of the different languages of research. He could determine the burden, spread, treatment, and cost of tuberculosis because he combined the talents of a demographer and an epidemiologist, a biologist and a doctor, an economist and a policy expert. "There was incredible expertise in the world on malaria or tuberculosis or any other given disease or problem, but there was nobody standing back and saying, 'What is the landscape?'" he realized. "If you don't have the big picture, it's incredibly easy for groupthink to lead you to focus on a limited number of things and you might miss what's really important."

By now, people at the Pop Center knew that even by Harvard's exalted standards, Chris Murray was extraordinarily productive.

In 1991, he received his MD and advanced from research fellow to assistant professor in the School of Public Health. He also began a residency in internal medicine at one of the nation's premier teaching hospitals, Brigham and Women's Hospital in Boston. The Brigham, as it is called by those who work there, had a unique "hemi-doc" program, allowing select doctors with existing research projects to split time between their laboratory or office and the wards. On a month, off a month became Murray's medical schedule.

Murray was already collaborating with Barry Bloom, then a TB expert at the Albert Einstein College of Medicine in New York and adviser to the WHO. "He was an assistant professor at the Harvard School of Public Health and a full-time medical school student at Harvard Medical School," Bloom remembers. "I would call him at three o'clock in the morning when he was on the wards. That was the only time you could get him. We could talk about TB policy and what needed to be done."

Together Bloom and Murray published an article in *Science* on the emerging threat of tuberculosis—not least the nearly 20,000 infected individuals currently living in New York City. "TB was completely off the agenda here," says Bloom. "It was terrible." Congressional hearings and new NIH research funding followed. In big cities, workers were hired to screen and treat people for the disease, particularly high-risk groups like intravenous drug users and low-income immigrants. It still wasn't enough, Bloom said, but at least "TB got a fair shake that it would not have otherwise."

Soon Murray gained a new sponsor that would enable him to take his TB research worldwide. The World Bank wanted help designing a large loan for tuberculosis control in China, and it sent him there to visit specific provincial health programs. On Murray's recommendation, the World Bank decided to put $50 million into TB projects in China—a huge investment in 1990.

The money bought new diagnostic equipment, training in proper lab work for hospital and clinic staff, plus modern drugs for the disease. It was now much more likely if you showed up with tuberculosis, you would be diagnosed, put on drugs, and the treatment would work.

"That was incredibly satisfying," Murray would remember. "Here was a disease that was really, in a global health sense, neglected. It had fallen off the radar in the '80s. There was no interest in it and the perception was there was nothing you could do about it in the developing world."

In 1992, two years after Murray and Styblo published their first article together and one year after Murray got his MD, the World Health Organization formed a new steering committee on tuberculosis research and made Murray, the recently appointed Harvard assistant professor and medical resident, its chair. Within three years, the WHO would endorse short-course therapy as one of its top global disease control strategies, a shift that the organization estimates has saved more than five million lives, one third of them women of childbearing age and children.

By framing Styblo's work in a way that public health researchers and policy makers could no longer ignore, "you could actually see things change," Murray says. "That made a huge impression on me—the idea of figuring out what's important and being able to communicate the results."

As a follow-up, Murray co-wrote books on adult mortality in China and the health of adults in the developing world. "Nearly 90 percent of children in developing countries survive to be adults, even in some of the poorest countries of Sub-Saharan Africa," noted the introduction to the latter book. How many ten-year-olds from eastern Niger became victims of heart disease or lung disease, cirrhosis or hepatitis? Murray wondered. How many ended up injured by a car accident? How many didn't have enough to eat? How many would later suffer giving birth? How many would cough blood? What he was doing was still unique. "He was one of the first people who started to think that mortality and health actually mattered for adults," recalls Alan Lopez. "This was pretty brave." As late as 1993, a journal article Murray co-authored could be considered provocative for the title, "Adult Health: A Legitimate Concern for Developing Countries." "People didn't care about it," Lopez says. "They didn't measure it. They only knew about child health."

Lopez understood from experience. With his supervisor at the WHO, he had tried to tally adult deaths by cause across the whole spectrum of diseases. There were no other estimates available. There was nothing with which to compare his numbers. "This was pre-HIV/AIDS," Lopez remembers. "There were no survey programs. No one was improving vital registration. There was just systematic neglect." Even the different disease control programs at the WHO did not make global estimates of adult mortality from the disease that was their specialty.

Looking at all available records circa the year 1985, Lopez estimated that every year 15 million children died, but so did 15 million people between age fifteen and sixty. These were premature deaths as well—tragedies, not inevitabilities. Think of a teenager who dies in a car accident, a twenty-something who commits suicide, a father killed by a heart attack in his forties, or a mother taken by breast cancer in her fifties. "One of the things that was coming out of my numbers and surprising me was the vast numbers of young adult deaths that were not being counted," Lopez would recall. "What I want to show," he decided, "is the importance of not just keeping children alive to adulthood, but adults to old age."

Tobacco would become for Lopez what TB was for Murray: a vast killer ignored in developing countries, in part because no one had comprehensively quantified its toll. Collaborating with Richard Peto, an Oxford medical statistician and epidemiologist, Lopez concluded that 1.5 to 2 million people died annually from smoking-related causes in poor- and middle-income countries. Their numbers were growing, yet these people had virtually no access to prevention efforts or chronic disease treatments. They had the same cancer, emphysema, heart disease, and stroke as smokers in rich countries, that is. What they didn't have were oncologists, pulmonologists, and cardiologists. Only TB claimed more adult victims.

Because adults died from different causes than infants, extending their lives generally required completely different strategies. A young boy given oral rehydration therapy to treat diarrheal diseases could be killed a few years later by HIV/AIDS. A girl vaccinated against measles had no protection as a young adult against rheumatic heart disease or cervical cancer. Wealthy nations like the United States or Australia had well-established injury and risk prevention programs. But because no one measured them, fires, falls, drowning, poisoning, road injuries, and other accidents weren't recognized as important causes of death in poor countries. Few imagined that, taken together, injuries could be a killer on par with diseases.

In the early 1990s, Lopez chaired a committee of leading demographers studying adult mortality for the International Union for the Scientific Study of Population. He couldn't convince them to add Murray, with his medical training, as a member. "Demographers stop just at mortality," Lopez would explain. "They don't travel into the causes of mortality. That's where Chris came in. I was trying to move them into that no-man's-land. I was subtly raising the importance of preventing adult mortality from heart disease, from murders, from accidents, from cancers."

It wasn't his only battle. Lopez was now editor of *World Health Statistics*, the WHO's compilation of health-related data for each of its member states. To make clear the absurdity of relying on national reports alone, he started including estimates from other UN agencies alongside the very different figures countries themselves had provided. These weren't as good as the numbers he'd come up with, he thought, but that was a separate battle. The estimates' virtue was that they were already "official," and so could be printed. In response, he encountered a pattern of quiet but still forceful political and scientific resistance. "I was getting more and more unpopular at the WHO," Lopez remembers. "I was taking a stand."

Murray meanwhile performed his hospital rounds, pondering ways to measure how we live as well as how we die. On the days he was on call, he arrived at the Brigham at 5:30 a.m. He did a solo round, talking to patients, checking labs. He noted any changes overnight. Depending on his rotation, he might see men and women with pancreatitis, heart failure, hepatitis, diverticulitis, or infections related to shooting up drugs. TB, especially in immigrants or people who'd been abroad. COPD (chronic obstructive pulmonary disease), often with pneumonia. Severe complications from diabetes. Each patient was different, but all were scared, scarred, or in pain. Their loved ones were terrified, too, facing the possibility of heartbreaking loss. Often they had to make decisions when neither they nor the doctors were really sure what was best. "Medical training gives you an insight into the nature of disease, but also into how this afflicts people," Murray would say later.

Residency, with its long hours, confined quarters, and constant stress, narrows a doctor's world to a few intense friendships. "It's like the army," Murray liked to say. His best friends in the wards were Jim Yong Kim and Paul Farmer, physicians and anthropologists who'd started a nonprofit organization called Partners in Health to provide American-level medical care—a clinic, hospital, and community health system—in Haiti. Their work would spread worldwide and both men would become celebrated global health leaders, but in the early 1990s all that was in the future. When not seeing patients, the three earnest and dedicated young workaholics debated what they saw as the most important question they faced: How to improve the lives of the world's poorest people—people seemingly ignored by everyone else? "The people we worked with, their idea of a very broad question was about the patient and the family," remembers Farmer, now chair of the Department of Global Health and Social Medicine at Harvard Medical School and the chief in the Division of Global Health Equity at the Brigham. "Chris's idea of a broad question was the political economy of health and wellness in the world."

In stolen moments between the Bunker at the Pop Center, his and Agnes's small apartment, and the wards at the Brigham, Murray asked himself what would it mean to treat the entire world as your patient. The health landscape was constantly changing: save the lives of children and they took on new risks as teenagers; help teenagers and they soon required the same services as middle-aged adults; treat middle-aged adults and it was among the elderly where new efforts would have to concentrate. Meanwhile, a disease like HIV/AIDS could appear at any moment, spread quickly, combine with a lingering malady like tuberculosis, and devastate populations worldwide. What was necessary was a system of measurement as dynamic as illness itself. How can we examine *everything*, Murray wanted to know, so we can know what's really wrong with people, and cure them?

Imagine that we could all agree on one definition of the impact of every illness or injury, Murray thought. Imagine this definition were so exact it was actually an equation, the sum of everything bad that ever happened to us from the day we were born until the day we died. All of a sudden, we could weigh health problems against each other: this many bouts of asthma on one side, for example, this many broken legs on the other. We could say *This problem is this big. That problem is that big.* On the exact same scale. "Obviously, being in good health is avoiding dying," he understood, "but it's also being able to move around well, being able to see and hear, being able to think clearly, and not being in pain, not suffering from anxiety, and not being depressed. It's common sense. These things really matter to how you live your life. But if you just focus on death, you miss them."

Not everyone agreed these were important issues. The scope of adult suffering was arguably even lower on the research radar than the spread of TB. In 1980, only a decade earlier, James Fries, a professor of medicine at Stanford, had published one of the most widely cited public health papers ever, entitled "Aging, Natural Death, and the Compression of Morbidity." By "compression of morbidity," Fries meant a decrease in the amount of time in life people spent ill. In societies everywhere, he argued, how *well* people lived was improving much faster than how *long* they lived. "Despite a great change in average life expectancy," he wrote, "there has been no detectable change in the number of people living longer than 100 years or in the maximum age of persons dying in a given year."

The race of life had a set maximum distance, in other words, according to Fries. But changes in lifestyle could shorten how long you spent panting before the finish. Whereas others anticipated an "ever older, ever more feeble, and ever more expensive-to-care-for populace," he wrote, his analysis suggested that "the number of very old persons will not increase, that the average period of diminished physical vigor will decrease, that chronic disease will occupy a smaller proportion of the typical life span, and that the need for medical care in later life will decrease."

Nonsense, Murray thought. As he saw every day at the Brigham, people could stay sick or injured without dying for decades in industrialized countries, particularly as new treatments addressed heart disease and stroke. What ailed us could be as bad—or worse—than what killed us. And everyday pain and suffering surely led to the majority of health care visits and money spent. Yet leading figures throughout world health policy circles were citing Fries, saying, "People are getting healthier and they're dying at the same age, so they're getting sick and disabled for less time." The implications for health services were huge. Did Americans need more or fewer cancer wards, diagnostic equipment, drugs, and follow-up care? Should the National Institutes of Health pour billions of dollars into trying to find a cure for diabetes, or would most people die before they needed it? Should new doctors train in pediatrics or in gerontology? In surgery or physical therapy? It was all a guess.

If what was happening to relatively rich adults, in a country with a complete vital registration system, was so little understood, consider ignorance about everywhere else in the world. How bad was autism in El Salvador, asthma in Iraq, clinical depression in China, or liver cancer in Zimbabwe? Well, we had no idea. People with these conditions were abandoned to their own fate. Their pain and suffering would not be recognized in any official statistics. And without official recognition, successful treatment was almost impossible. We have to track them, too, Murray thought. But how?

CHAPTER FIVE

The Big Picture

Quantity and quality—Traffic versus cigarettes—The
assignment.

Everyone on Earth should have the chance to live a long life in
full health. But the statistics we used to measure progress toward that goal were not only inaccurate, but also often irrelevant.
What was needed was a single measure of how diseases and conditions cost us both quantity and quality of life.

By 1990, both Chris Murray, at Harvard, and Alan Lopez, at
the WHO, had found a powerful outside ally. Dean Jamison, former chief of the World Bank's Population, Health, and Nutrition
Division, was leading a comprehensive review of disease control
priorities in developing countries at the University of California
at Los Angeles. Was nutritious food most important, or access to
TB drugs? AIDS treatment, or accident prevention campaigns?
What policy investments, Jamison wanted to know, produced the
greatest public good?

The answer to that question depended on the person responding. Predictably, individuals spoke up for the policies on which
they worked. *Clean water!* Jamison heard one day. *Antimalarial
bed nets!* another. *Breast-feeding! Diet! Safe childbirth! Vaccines!*
Meanwhile, even the best-intentioned people missed important
health problems just because they were nonfatal or had no one

counting them. Whenever advocates of this or that program promised their specific intervention was the most important, Jamison liked to ask what program was *second-most important*? No one could answer him. No one seemed to know why his or her area was more important than any other, or even what the others were. No one seemed to want to know.

Without hard numbers, comparison was impossible. Without comparison, any claim for priority was one person's word against another's. "If you're thinking of spending money on measles immunization or polio immunization, measles kills people, but, if they survive, they're okay," Jamison put it. "Polio kills fewer people, but a lot of people have a lot of disability if they survive." So what was more important? "If you're going to make this decision between one vaccination or another," Jamison told people, "you have to be de facto trading off."

Activists and aid workers have an understandable response to this kind of lecture: so-called prioritization kills. We need more money for everyone, they say, so no one is left without care. And events would soon show far more money was available for international health aid than anyone had imagined. Yet the problem Jamison was identifying applies to even the wealthiest and most progressive health systems. What is the total health loss from each disease and injury? Where are those losses concentrated, so we know where to focus prevention efforts as well as cures?

With Jamison, Murray reviewed various attempts since the 1960s to create a combined measure of impairment, illness, and death. Look at life span alone and the devoted gardener who never has so much as a cough and dies at age eighty seems the same as a neighbor who also lives to age eighty, but is blind, bedridden, or paralyzed by anxiety attacks. Simply count deaths, on the other hand, and the pneumonia that kills a one-year-old is no different than the stroke that kills a seventy-year-old. To know the total health loss from any problem—"the burden of disease," he and Jamison started calling it—Murray concluded that you had to measure *the years of healthy life lost*, not just the ages at which people died or the number of lives taken.

Murray defined burden with a two-part sum. The first part was about everything that kills people. Say, as was true in 1990, that men and women in the healthiest places on Earth could expect to live about eighty years on average. Then if you died at any age short of eighty years you had "lost" that many years of life—at least compared with the ideal. Demographers called this shortfall "potential years of life lost" (YLLs). For example, if a stroke killed you at age seventy, you had lost ten years of potential life to early death. If pneumonia killed you on your first birthday, you had lost seventy-nine years. In terms of potential years of life lost, then, the case of childhood pneumonia was almost eight times worse than the stroke.

- -

Sample Calculation of Years of Healthy Life Lost to Early Death

- -

CAUSE OF DEATH	AGE AT DEATH	YEARS OF HEALTHY LIFE LOST (ASSUMING IDEAL LIFE SPAN OF 80)
Stroke	70	10
Pneumonia	1	79

- -

The second part of the sum concerned nonfatal health problems, weighting each on a scale from 0 (perfect health) to 1 (death). If you thought of deafness as being a fifth less healthy than perfect health—0.2 on the 0-to-1 scale—then living a year without hearing could be thought of as equivalent to losing 0.2

years of healthy life. In the same manner, having mild neck pain might be roughly equivalent to a tenth less healthy than perfect health (0.1 on the scale), and having untreated major depression might be six tenths less healthy (0.6 on the scale)—so each year lived with mild neck pain would mean 0.1 years of healthy life lost, and each year lived with severe depression would be 0.6 years of healthy life lost. Echoing the demographers' "years of life lost" (YLLs), Murray called this new statistic "years lived with disability" (YLDs).

Sample Calculation of Years of Healthy Life Lost to Disability

---

CONDITION	DISABILITY WEIGHT	YEARS OF HEALTHY LIFE LOST PER DECADE WITH CONDITION
Mild neck pain	0.1	1
Deafness	0.2	2
Severe depression	0.6	6

Comparing having various illnesses and disabilities with dying early would certainly be controversial—how, for example, would the exact weighting for each condition be chosen?—but the idea was also refreshingly comprehensive. Rather than ignoring the health loss from back pain, or blindness, or bipolar disorder, or cancer treatment, the weighting promised for the first time that disability would be counted seriously. Take the example of being hit by a car while crossing the street. "If someone dies of cancer at age seventy-five, maybe the disease has taken five years of life," Murray said. "If he dies of a car crash at age twenty-five, though, that's taken fifty-five years. And if he survives the car crash, but has a severe spinal injury, and then dies at age sixty, that's both twenty years of life lost and thirty-five intervening years lived with disability." All of a sudden, traffic might be worse for you than cigarettes.

Sample Calculation of Total Years of Healthy Life Lost

Fatal cancer at age 75

75 years of healthy life   5 years of
                                                  healthy life
                                                  lost to early
                                                  death

Fatal car crash at age 25

25 years of healthy life   55 years of healthy life lost to early death

Car crash at age 25, severe spinal injury (disability weight 0.6), death at age 60

25 years of healthy life   35 years of injury with a disability weight   20 years of healthy
                             of 0.6 per year; 35 x 0.6 = 21 years of   life lost to early
                             healthy life lost to disability   death

---

The elegance of this new formulation for death and infirmity was that both parts of the sum shared the same unit: years of healthy life lost. Add them up, and you had what Murray termed the number of disability-adjusted life years, or DALYs, attributable to any health problem. In plain English, the number of disability-adjusted life years (DALYs) was the sum of years of life lost because of early death (YLLs) and equivalent years of life lived with disability (YLDs). Expressed as an equation, DALYs = YLLs + YLDs. Quantity of life plus quality of life lost.

DALYs, aptly enough, rhymed with tallies, and the metric, if calculated for whole nations, could allow all manner of comparison: not only between what killed people and what merely made them sick, but also between where and when and whom was hurt—all in a single number. Was measles worse, or polio? AIDS or osteoarthritis? Drug abuse or alcohol? Was personal health better in Latin America or Eastern Europe? Canada or South Korea? For five-year-olds or fifty-five-year-olds? For women or for men? Count total DALYs by cause, or DALYs per capita by region, age group, or sex, and you could finally see. In this way, the new measure would be the health equivalent of gross domestic product, which purports to total every element of a national economy from bagels to battleships. Except, in general, countries want their economies to be as large as possible, whereas they would want their total health loss to be as small as possible. Both measurement and benchmark, DALYs would summarize all the health problems for everyone, at every age, everywhere. For the first time, we could see everything wrong with us.

"Chris," Alan Lopez would say, "converted me from death to the importance of also measuring the impact of morbidity and disability on populations. That," he realized, "is an extraordinarily beautiful and extraordinarily useful policy tool."

Murray's new concept, DALYs, was a radical invention that would be refined and improved many times in the following years. But the basic idea had come from a simple and powerful shifting of perspective. Like the recognition of the 10/90 mismatch between research funding and health needs and the earlier realization that global mortality statistics were grossly unreliable, DALYs came from the desire to stop seeing international public health in fragments and find a single view of everything. The result would be a set of measurements as potentially transformative as the first maps of the world produced by the Age of Exploration. In the sixteenth century, for the first time, you could look at a map and see the Americas as well as Europe, Africa, and Asia. Coast and ocean boundaries were reasonably realistic. Latitude told you how far you were from the equator. Atlases and Mercator projections brought the whole world together in one book or even on a single page. Devising a method to chart the same global picture for how all humanity lived and died had taken another four hundred years.

In 1991, the World Bank decided to devote the 1993 issue of its annual flagship publication, the *World Development Report*, to investing in health. Dean Jamison, chosen as the report's editor in chief, turned to Murray and Lopez. DALYs should be central to the analysis, he suggested. They could help identify the most significant priorities for health spending in a way that would clear away much of the prior confusion from competing claims. They called the exercise to measure DALYs from all causes for all people and places the Global Burden of Disease study.

Professional etiquette required that Jamison interview others before offering Murray, still only twenty-nine years old, the position of burden team leader. But he knew that with a hard deadline less than two years away and Lopez constrained by his position within the WHO, there was no other choice than the first-year medical resident and his new metric. Besides, Murray had been contributing to disease control priorities research for four years. "All the ideas were out there," Jamison would remember later. "Our job was to take the best we could find and make them work."

For Murray and Lopez, it was the assignment they had been working toward for a decade—a once-in-a-career opportunity to create a whole new science of health measurement from scratch with guaranteed impact on the ground. "Count up health problems by disease," Murray described the assigned task of the *World Development Report* researchers, "count up how much you can change health problems by specific investments, and you have the big picture and strategies to improve health." It was his and Lopez's chance, at long last, to help everyone.

Taking such a prominent position in health policy priority-setting was new terrain for the World Bank, which was much better known for its economic expertise and loans for infrastructure. As for why it was now interested, the answer was simple: All the economic development in the world meant nothing if people weren't healthy. "Look at the United States," says Larry Summers, then the World Bank's chief economist. "Would we rather have a 1900s standard of living and today's health care, or 1900s health care and today's standard of living? Most people would say 1900s standard of living and today's health care. That's saying this relatively limited sector of the economy has done more good than all economic growth in all the other sectors of the economy."

In the past, World Bank staff had argued that markets ensured social welfare better than big government interventions. In delivering the greatest health at the lowest cost, though, free market capitalism had often failed. The people who were most sick were generally also those least able to pay for medical treatment. Without first being treated, however, they had little chance of making money. "Because good health increases the economic productivity of individuals and the economic growth rate of countries, investing in health is one means of accelerating development," the World Bank would conclude. "More important, good health is a goal in itself."

Part of the good of getting government out of doing everything was that this freed it to put more energy into what it *was* good at. Health care, for example. Despite many real shortcomings, national and international health programs could be extremely effective—including cost-effective—at providing medical care to entire populations. And they were even better at the preventive measures, from clean water to vaccinations to antismoking campaigns, that went under the name of public health. Now Larry Summers wanted the World Bank to issue clear calls to action. "I had become tired of what I regarded as platitudinous Bank prose, which constantly said, 'This is an important area. Policies must take this into account,'" he recalls. The upcoming *World Development Report*, Summers told Dean Jamison and his staff, better "have some bite."

Murray and Lopez were ready and eager. The Global Burden of Disease study, Murray promised Jamison and Summers, would deliver. This new way of measuring health, if it could actually be accomplished, would uncover some surprises and many revelations—particularly when done on a worldwide scale.

PART II

What Doesn't Kill You

CHAPTER SIX

A Global Checkup

Balancing acts—"How can *you* say anything?"—Punters—
Nails.

People in the tech world like to talk about "granularity" and
speak approvingly of a "granular" database. What they mean
is a huge field of information that can be zoomed out to show a
really big picture or viewed up close in such detail it's like looking
at an individual grain of sand. In recent years, millions of us have
become familiar with moving dramatically from macro to micro
through programs like Google Earth, where you can survey entire
continents or focus down to a single cat sitting on a windowsill.
In the 1980s and early '90s, when Chris Murray and Alan Lopez
started their effort to chart world health in minute detail, such
granularity was not yet available in epidemiology, nor had anyone
devised a method of achieving it. All they knew was that to do it
all and not miss anything, you had to be systematic.

Murray, the doctor who wanted the world to see the big picture
of public health, still spent much of his time on the ills of individual
patients at the Brigham. With the attending physicians, he devised
treatment plans for the day: put in arterial lines, tap fluid in a patient's lung, do blood cultures, schedule MRIs. In the grueling tradition of medical residency, a typical shift lasted forty hours. Murray
left between 6 and 8 p.m. He was due back in less than twelve hours.

Rather than sleep or go home, he traded his white coat for a black suit and tie, the off-duty dress code for a Boston doctor, and trekked across the Charles River to the Pop Center annex in Cambridge. "I don't know," he says when asked how. "I had to be there." Sometimes he was so tired that he forgot to remove his stethoscope and arrived with it still around his neck. But there was no way he could use it on all the patients who now awaited him. His role in the *World Development Report* required the young resident to give 5.3 billion people a complete physical and psychological review.

Lopez, at the World Health Organization, had his own balancing act to perform. From the outset, the WHO was opposed to his work on the *World Development Report*. The organization had gone along, he thought, only to keep the World Bank from undermining the WHO's mandate, written into its constitution, to lead the world in health measurement. Seeking greater independence, Lopez arranged to move from the health statistics division to a newly formed WHO tobacco control program, where he would be responsible for estimating the health effects of smoking worldwide. "Ostensibly, on the outside, I started to work on tobacco, which I enjoyed," he remembers. "But I was also, in parallel, starting the big work on the global burden of disease." With the support of his wife, Lene, who stayed in Geneva with their young daughter, Inez, Lopez flew across the Atlantic on weekends, holidays, and whenever else he could justify it at work, meeting with Murray at the Pop Center in person to share data and consult on methods. "We had to understand global mortality if we were going to understand tobacco mortality," he says, using the same logic he employed to get permission for his many trips.

To begin, Murray and Lopez studied the WHO's *International Classification of Diseases* and made a list from it of every major world health problem, from HIV to polio, drug dependence to glaucoma, iodine deficiency to war injuries. They settled on about

100 diseases and injuries accounting for almost all deaths as well as what they estimated was more than 90 percent of the global burden of disease attributable to disability. Because the ultimate goal of the *World Development Report* was improving policy, they organized these maladies in three broad existing categories by which health services and public health efforts addressed them.

In Group I were what almost everyone thought of when they considered the concerns of poor countries: communicable diseases contracted by contagion or infection and health problems related to being born or giving birth. In Group II were commonly considered "rich country" problems: noncommunicable diseases such as cancers, addictions, heart disease, and depression. In Group III were injuries, intentional and unintentional—poisoning, drowning, road accidents, suicide, and other grisly events—which, at the time, virtually no one studied on a global basis.

Lopez concentrated on mortality and cause of death. In his work in Geneva over the previous decade he had established several new databases for the WHO based on the application of the *International Classification of Diseases* by countries. These databases allowed him to classify deaths consistently from cause to cause, region to region, and between age groups and men and women. For his expertise and all-consuming interest in mortality, colleagues had nicknamed him "Dr. Death."

As Lopez had already discovered, a major challenge was that physicians classified the same kind of case very differently from country to country or even from year to year. The percentage of French deaths said to be due to cancer, for instance, was 10 percent higher than it would have been if the reporting standards of the United States had been applied. What seemed a 5 percent higher death rate, in other words, was in fact almost 5 percent *lower*. With his scientific training and his experience, Lopez would look at the information and say, "This is the data source—let me try a different approach and internal consistency check." Place by place, disease by disease, study by study, he tried to reconcile disparate numbers.

Initial Global Burden of Disease Categories

GROUP I	GROUP II	GROUP III
Communicable diseases and maternal and perinatal conditions	Noncommunicable diseases	Injuries

SAMPLE CAUSES	SAMPLE CAUSES	SAMPLE CAUSES
Dysentery	Alcohol dependence	Drowning
HIV	Depressive disorders	Falls
Hookworm	Diabetes mellitus	Fires
Leprosy	Glaucoma	Homicide and violence
Malaria	Ischemic heart disease	Motor vehicle injuries
Maternal hemorrhage	Leukemia	Occupational injuries
Measles	Parkinson's disease	Poisoning
Schistosomiasis	Rheumatoid arthritis	Self-inflicted injuries
Tuberculosis	Stomach cancer	War

More vexing were what he and Murray termed "garbage codes"—cases where the recorded cause of death makes no medical sense or is so general as to be of little value. No one ever really dies of "senility," for example, though that is often listed as an official cause of death. Even "brain trauma" is too vague—a fatal brain injury could derive from a car crash in one case and from falling in another. You have to know which to prevent future fatalities. To Murray and Lopez, a doctor who coded an underlying cause of death as "brain trauma" was like someone calling the fire department and telling the engine team to look for the house on fire rather than where it was. Yet garbage codes were—and are—extremely common. "Heart failure," "liver failure," "pulmonary embolism"—in some areas, *more than 40 percent* of a country's official cause-of-death claims can be garbage. As Lopez worked, all garbage codes required reassignment based on the most likely true underlying cause.

For people living in countries without any real death registration data—in the worst case, close to 98 percent of sub-Saharan Africans—Lopez and Murray constructed new statistical models to prevent double counting and predict different causes of death for men and women as they aged. You had to be very careful, however. A simple model might say Australia had this many children per thousand born with Down syndrome, so Angola, where data was sketchy, would, too. A more robust model would factor in things like the age of the parents—or tell you that Australia and Angola were too different to compare, and that you needed numbers from South Africa instead. The Global Burden team produced dozens—sometimes hundreds—of models for any area of interest, and calculated which ones best matched special cases where they had good real-world data.

At the same time, Murray and a small group of Harvard undergraduates, recent graduates, and grad students tracked disability—Global Burden's official term for any nonfatal health loss. Using a custom computer program Murray had designed, they listed key consequences that might occur from each disease and injury included in the study. For example, diabetes mellitus could lead to vision loss, nerve damage, or lower leg amputation. A traffic accident could produce severe burns, a fractured skull, an eye injury, or a dislocated shoulder, to name just four of thirty-three outcomes the team identified from that one cause. What data they found worldwide suggested how often each of these consequences occurred both overall and by sex and age. Six percent of very young children in road accidents appeared to suffer open wounds, for instance. Almost a quarter of those over sixty in the same circumstances had a fractured femur, rib, or sternum. The figures were not nearly as precise as they would become in the future, but the gaps in the big picture were beginning to be filled.

The *World Development Report* would divide the globe into eight demographic regions: sub-Saharan Africa, India, China, other Asian countries and islands, Latin America and the Caribbean, the Middle Eastern crescent, formerly socialist economies of Europe, and established market economies. In any given year, from a region's general population, some percentage suffered a particular disease or injury. Those with the condition either got better, returning to the general pool, or they developed some disabling consequence, or they died. A weighted measure of resulting time spent disabled would equal years lived with disability—the YLD component of Murray and Lopez's calculations.

How, though, to compare having these different disabilities with each other and with dying early—how, that is, to weight each condition on Murray's 0-to-1 scale, 0 being equivalent to perfect health and 1 being equivalent to death? To outsiders, this was perhaps the most controversial and audacious part of the entire Global Burden study. But what had seemed to some an impossibly subjective task actually generated considerable agreement when Murray convened independent panels of experts in international health to make the evaluations. Eventually, these panels would expand to include members of the general public worldwide, but the level of consensus remained the same. As future studies would show, there was a clear common perspective on the severity of different conditions. And it could be stated as a number, which was necessary if the leading causes of pain and suffering were to be compared with one another and with the major causes of death.

Early Burden of Disease Disability Severity Weighting

SEVERITY WEIGHTS   SAMPLE CONDITIONS

0.00–0.02	Vitiligo on face
0.02–0.12	Watery diarrhea, severe sore throat, severe anemia
0.12–0.24	Radius fracture in a stiff cast, infertility, rheumatoid arthritis
0.24–0.36	Below-the-knee amputation, deafness
0.36–0.50	Rectovaginal fistula, mild mental retardation, Down syndrome
0.50–0.70	Unipolar major depression, blindness, paraplegia
0.70–1.00	Active psychosis, dementia, severe migraine, quadriplegia

---

At last, the team was closing in on the single view Murray had outlined of the entirety of human health loss. For every person, take the years of life lost to an early death and the equivalent years of healthy life lost to disability, and you had the burden of disease—by cause, sex, age, and location. But how would the problems people actually faced match up with the remedies health systems and international public health campaigns offered?

In December 1992, four months before the Global Burden portion of the *World Development Report* was due, Murray and

Lopez presented the first set of preliminary results at an all-day meeting at the WHO in Geneva. As seemed to be their habit, the two had worked through the night, finishing at four in the morning, accompanied only by Giles, the Lopez family West Highland white terrier. "This is pre-PowerPoint," Murray remembers. "The meeting started, and we had to ask the chair to stall while the secretary went off to make transparencies." Every other floor of WHO headquarters includes a double-height meeting room, called a *salle*. The one assigned them was packed as department heads and international staff members in each program area gathered here from all wings of the building.

The very idea of Global Burden—counting everything for everyone, everywhere, including disability as well as death, and ensuring total deaths by cause did not exceed total mortality—was new even to the many public health experts in the *salle*. In the discussion that followed, though, it became clear that the possible merit of this novel approach struck many as secondary to a more pressing issue. People were most concerned about how this way of measuring things might adversely affect their own work. "Member nations, scientists within the WHO, scientists outside the WHO—all of these had something to lose," Lopez observes. Overestimating how bad a disease toll was could lead to more money and attention for your cause or country. Underestimating the count could suggest that you had made great progress and deserved increased funding or another term in office. Aid groups each wanted their issue to be recognized as the most pressing. Researchers wanted their subject matter to be recognized as the most important. No one compelled different groups to make sure their numbers were comparable with each other.

The very structure of operations in Geneva was an unintended illustration of the dangers of working without the big picture. At WHO headquarters, well over 90 percent of staffing and resources were devoted to communicable diseases and the problems of pregnancy, childbirth, and early childhood. These were vitally important, but they accounted for less than half of total health loss, according to Murray and Lopez. Almost as large—42 versus 46 percent—was the burden from noncommunicable diseases. And injuries were 12 percent—a huge proportion given that the entire WHO injury prevention program still had just one staffer.

Breaking the data down by sex, age, and region revealed even more alarming gaps between public health needs and the current allocation of resources. In developing countries, Murray and Lopez said, motor vehicle injuries were the third-leading contributor to health loss for young adult men. Depression was the fifth-leading contributor to health loss for young adult women. Suicide, for both sexes, was sixth. Osteoarthritis—a disease that killed no one—was ninth. In sub-Saharan Africa, they estimated, the overall toll from *dental problems* equaled that from anemia. Throughout Asia, ischemic heart disease took more healthy life years than complications from pregnancy, and neuropsychiatric diseases hurt more than nutritional deficiencies. And in the Middle Eastern crescent, health problems from injuries were four times worse than those from cancer. All of the above were surprises to the Global Burden team—and shocks to the status quo.

If Murray and Lopez were right, the World Health Organization's gigantic global outlay of time, money, and passion was doing next to nothing about nearly half the health problems of the world. People who decided on priorities at the WHO, and in countries and donor agencies, were being told that they had miscounted their patients, misrepresented their progress, and just plain missed serious health problems happening before their eyes.

So, not surprisingly, their first response was to argue that Murray and Lopez were wrong.

"Generally, the reaction was: 'I'm person X, I'm in territory A, how do you know N?'" Murray would recall. "The standard was 'We don't know. We don't have good enough data. So how can you say anything?'"

Because we have to, Murray and Lopez answered. These were not problems that could wait for perfect data. "Choices between competing health priorities are made every day by decision-makers in the public and private sectors," they would write. "These choices reflect each decision-maker's implicit understanding of a population's epidemiological profile, as well as opportunities for intervention. We believe that it is preferable to make an informed estimate of disability due to a particular condition than to have no estimate at all." Better to push epidemiology to its limits—and then improve it—than to hold back, waiting for more information, and let the pain and suffering of billions remain ignored.

Their day in Geneva was exciting for Murray and Lopez, even with the pushback. They had engaged people. When Murray got back to Cambridge, he presented the same findings at a Pop Center seminar. "That was a pretty small group, but it was pretty intense," he would remember. Amartya Sen, the Harvard economist who would be awarded the 1998 Nobel Prize in Economics, was there. So was Arthur Kleinman, chair of the Department of Social Medicine at Harvard Medical School, Sudhir Anand, a prominent professor of economics at Oxford, and Sissela Bok, the philosopher and ethicist who was also the daughter of two Nobel Prize winners. Those in attendance closer to Murray's age included everyone from the future head of the World Bank's global HIV/AIDS program to the future deputy executive director of BRAC (formerly the Bangladesh Rehabilitation Assistance

Committee), the world's largest nongovernmental development organization in terms of both employees and people served. Each took a seat around a large oval-shaped antique wooden table.

Pop Center seminar tradition encouraged lively and even sharp exchanges among all the attendees, no matter what their distinction. "Nobody was ever made to feel that he wasn't asking the right question, or that you shouldn't talk if there was a Nobel Prize winner in the room," one regular remembers. "All of them had an interest in hearing everyone."

Again, there were objections, though now from different perspectives. Murray almost immediately found Sudhir Anand, the Oxford economics professor, and Kara Hanson, an up-and-coming health economist, questioning the assumptions underlying his calculations. Murray had grafted economic theory inappropriately on his health framework, they said. His central metric, disability-adjusted life years, was the grand total of years of healthy life lost from both early death and disability. But Murray had added a significant wrinkle. In what was called "age weighting," his final assessment of disease burden valued years lived in midlife greater than in childhood or old age, on the grounds that these were the period of one's greatest contribution to society.

This was ethically pernicious, Anand and Hanson argued, first at the seminar and later in print. "For example," the pair wrote in the *Journal of Health Economics*, "doctors' and nurses' time could be argued to be more valuable than that of other professions." Likewise CEOs might be said to contribute more to the world than the homeless, and the able-bodied more than the disabled. Murray countered by explaining that the weighted values for different age groups came from polling international health workers, but he admitted the issue required further consideration. In later versions, the weighting would be revised.

The central concept of DALYs—disability-adjusted life years—survived the criticism, although some were uncomfortable applying a new system of measurement to real-life policy recommendations. "On the DALY concept, there are people who are deeply concerned that somehow if you get the numbers wrong and you have a single metric, you will influence resource allocation in the wrong way," Murray would say later. But doing nothing, waiting for complete information or a consensus that might never come, was also a way of encouraging bad decisions. Look at the massive problems in everyone else's numbers that he and Lopez had discovered. Holding back on burden estimates was not reasonable caution or wise prudence—it was an evasion of responsibility.

"Think about your own family," Murray said. "You care about them. People care about them." You wouldn't wait until you knew for certain that your daughter had appendicitis before you took her to the doctor and noted that she was experiencing sharp pain on the right side of her abdomen. You wouldn't just go home in silence if the doctor refused to make a diagnosis. "In the research community, they say, 'We'll punt. We'll let decision makers put all this data together and make all the right decisions.' But decision makers have even less time and less ability to sift through all the necessary data."

Murray and Lopez didn't claim that their calculations were exactly right. Population health studies were never absolutely precise. They knew the limits of their data and their initial methods. But they also knew their estimates were comprehensive, they were impartial, and they tallied almost everything harmful to human health everywhere in the world in identical manner, using the best evidence available, for adults and children, women and men, poor and rich, what didn't kill you and what did. The most vulnerable people, globally and in every society, bore the greatest burden of disease and disability. Measuring that burden as fully as possible was perhaps the best way to demand that governments and international institutions improve their care.

Two weeks before the team's deadline, Larry Summers asked Dean Jamison the probability that the Global Burden project would be completed on time. Jamison paused. "Eighty-five percent," he said. Summers shook his head. "Not good enough," he responded. Jamison passed the message on to Murray, who told Lopez to join him right away for a final push.

From Geneva, Lopez flew to Boston's Logan Airport, landing at close to midnight before he made his way to the apartment Chris and Agnes rented in Cambridge, between Harvard and Porter Square, the closest place to campus the junior professor could afford. He and Murray rose at 5 a.m., in total darkness, to beat traffic to the Pop Center and then make the long drive to the snow-covered barn turned getaway office of a vacation house that Murray had bought for little money up in Maine. The license plate on Murray's tiny white Ford was "GBD1," as in Global Burden of Disease. The front seat was decorated with empty Dunkin' Donuts coffee cups.

As Murray sped forward, Lopez couldn't help but think back to what had brought him here. At age twelve, he'd left the quiet Western Australian country town of Narrogin where he'd grown up and entered Aquinas College, an elite boarding school in Perth, then a city of 500,000. His classmates were the sons of lawyers, doctors, real estate magnates, and other wealthy businessmen, while Lopez's father was a policeman who moonlighted as a school bus driver to make the extra money necessary for Alan's tuition. Lopez was always conscious of the sacrifices that had been made for his education. "My experience was very much affected by that. I needed to do well."

In his five years at Aquinas, Lopez was a top student in physics, chemistry, mathematics, and Latin, and helped lead his class's sports teams: cricket first term, hockey second, track third. Their coach, a Christian Brother, was so hard on the boys everyone called him "Nails." "You can do it!" Nails said. "Near enough is not good enough." "He got the best out of us," Lopez would remember later. "You would never go to complain to him." Try it, and he'd make you run five miles—and he'd run them with you.

It was perfect training for keeping up with Murray on the global burden section of the *World Development Report*, Lopez came to realize. "Twenty years later, when I meet up with Chris Murray, and we have this enormous task in front of us, I didn't say we couldn't do it," Lopez remembers. "I already had had it beaten into me that I could do it, we could do it, and we should do it, even if it was going to be hard."

Murray and Lopez holed up in the barn in Maine, compiling data in marathon sessions, 6 a.m. to midnight, in order to refine their calculations for final submission. The *World Development Report*, subtitled "Investing in Health," was published in June 1993. Woven throughout were burden-of-disease numbers, with the new concept of DALYs, disability-adjusted life years, introduced on page one. In eighteen months, with a budget of approximately $100,000, Murray, Lopez, and their collaborators had begun to transform our understanding of life and death.

Thank you, Nails.

Veritas Vincit, read the Aquinas College motto. Truth conquers.

CHAPTER SEVEN

Home and Away

Description and prescription—"Dear Christopher"—A
recruitment speech.

The 1993 *World Development Report* introduced a new way of measuring health as more than just the absence of death, and a new way of calculating the economic and social cost of disease and disability across the world. The next hurdle was getting policy makers and public health authorities to accept the change. That started with getting them to accept the numbers, which had been initial calculations from a fast-moving swirl of global data.

In 1994, the Harvard School of Public Health promoted Chris Murray from assistant to associate professor. He and Agnes had two young children now, Anne-Sophie and Timothy, and had bought a house in Acton, Massachusetts, twenty miles northwest of Cambridge, but Murray still worked almost nonstop. With fresh funding, and in collaboration with Alan Lopez, he formed an official burden-of-disease unit in the Bunker at the Pop Center, hiring several junior staffers to gather and refine ever more information about death, disease, and injury around the world.

Grouped together around a salvaged conference table, the team fed on takeout pizza and hundreds of pages of graphs. Peculiar patterns meant bad data or a surprising trend. The goal was figuring out which was which. Already, external pressure to shift or suppress the science was constant. The objections that had started at the WHO talk and Pop Center presentation grew louder. "Chris made some people enemies or at least very unhappy," remembers Catherine Michaud, a Swiss physician turned public health specialist, one of Murray's earliest collaborators at the Pop Center. "He would go to great lengths to explain his rationale and train of thought, but if people didn't agree with him, he didn't change his attitude. He wouldn't say, 'Oh, you are at the WHO. You are the authority on malaria. Therefore I have to revise my estimates to fit yours.' That was never the case."

People suffering from overlooked conditions took notice of their new champions and started using their data to bolster their standing. Advocates for heretofore neglected causes, particularly psychiatric treatment, injury prevention, and the relief of bone and joint pain, cited burden-of-disease findings to lobby for increased funding. The U.S. Senate called Murray to testify on diabetes, HIV/AIDS, and tobacco-related disease trends. When colleagues at Harvard edited a report on mental health problems in low-income countries and priorities for future care, figure 1 was a pie chart of global DALYs by cause, straight from the *World Development Report.* "This Report puts the issue of mental health and well-being firmly on the international agenda," wrote UN Secretary General Boutros Boutros-Ghali.

An even larger affirmation of the burden-of-disease approach came from Mexico. From 1992 to 1993, Julio Frenk, founding director of the National Institute of Public Health of Mexico, spent a sabbatical year at the Harvard Pop Center. There he met Murray and learned of the nascent Global Burden of Disease study. The importance of the work was obvious to him. "There was the realization, just as a doctor seeing a patient can't do an analysis without evidence-based medicine, we can't do adequate policy either within countries or at the global level without a foundation of scientifically derived evidence," Frenk remembers. "Here we have hundreds of diseases," he thought. "How do I know if TB is more important than cancer or not? What DALYs do is let me see the total burden of disease throughout the years, compare one disease to another, and make funding decisions."

Frenk sent his staff from Mexico City to Harvard to get training and perform a Mexico-specific burden-of-disease study. What they came up with became a 1994 book, a research project, and a nonprofit think tank center, all called "Health and the Economy." "Our idea was to make a proposal for politicians, like a menu of solutions to reform the health system," says Rafael Lozano, the team's head epidemiologist.

"Health and the Economy" was the first published use of the DALY concept after the 1993 *World Development Report* itself. Both works recognized that the burden of disease was really valuable to policy makers only if it was accompanied by a plan of action. " 'Sure,' politicians will say, 'we have a big burden from lung cancer,' " says Alan Lopez, " 'but what should we do about it?' " To that end, the *World Development Report* had specifically suggested that developing countries redirect government spending from specialized care to low-cost, highly effective immunization, hunger relief, and infectious disease control and treatment programs. These wouldn't solve every problem, but could reduce developing nations' burden by 25 percent on exactly the same budget.

Focused on just one country and written by insiders with access to much more data, "Health and the Economy" and its follow-ups did an even better job matching description with prescription. While Lozano, the epidemiologist, calculated the burden of disease across Mexico, José-Luis Bobadilla, a World Bank veteran, helped tally money spent on health, plus the price of different possible interventions and how much good each one did per peso spent. The reports that came out of "Health and the Economy" showed what made Mexicans sick, what they could spend to get better, and what treatments would make the biggest possible impact. "With that you can start producing proposals," says Lozano. "If you invest X, you will gain Y."

Julio Frenk oversaw the entire effort. "It completely changed the perception and sense of priorities in Mexico," he says. "Before DALYs, we assessed the importance of health problems by the number of deaths. Obviously, there are a lot of diseases that don't kill people, but produce a lot of disability. That's the case with mental illness." For the first time, not only depression, but osteoarthritis, arthritis, and low back pain were seen as vital state concerns, each nonfatal yet among the ten biggest burden-ranked health problems facing adults in Mexico. For men, and particularly young men, road traffic accidents cost many years of healthy life. For women, it was breast and cervical cancer. But few health programs in the country addressed depression, anxiety, joint pain, back pain, injuries, or cancers.

The Mexican health system was still built for the world of the early 1950s, when outreach efforts focused almost exclusively on promoting safe childbirth and preventing the spread of infectious diseases. In those days, a typical woman in Mexico gave birth seven times. At least 70 percent of children were born at home. Roughly one in six newborns died. The *average* age of death was twenty-four years old. "Surviving was not easy," Lozano says. "Vaccines didn't exist except for rabies and smallpox." Measles, mumps, and whooping cough killed thousands annually. Now, in the 1990s, enormous progress had been made in fighting communicable diseases. And women were having half as many children as in the 1950s. The trend showed clear convergence with the United States.*

Ranked by the number of deaths caused, according to "Health and the Economy," the leading health problem in Mexico in the early 1990s was cardiovascular disease. Ranked by DALYs, it was unintentional injuries. Only ranked by YLLs, or years of life lost to early death, were perinatal conditions worst, and then only because of especially adverse figures from the country's rural southern region. The national burden study, recalls Frenk, "totally changed the policy conversation." "It's not just about diarrheas," he says people realized. "There's a double burden"—communicable diseases *and* chronic conditions. "It's a much more complex picture."

Soon Frenk and Murray would have a chance to help lead the response—not just in Mexico, but around the world.

In 1996, Murray and Lopez published "Evidence-Based Health Policy—Lessons from the Global Burden of Disease Study," an article in the journal *Science* cited by more than 1,800 subsequent scholars. At the same time, they prepared a summary of the team's revised findings and methods for peer review. "I'd been trained by my father to submit to *The Lancet*," Murray would remember. So he did.

The Lancet's influence and importance had only grown since the 1970s. And it was even harder to get published there. If you were anyone in international health—from humble researcher to the director-general of the World Health Organization—*The Lancet* was the principal arbiter of your professional reputation. "Dear Christopher (if I may)," wrote back the journal's new editor in chief,

---

*Indeed, in 2010, Mexico would average 2.1 births per woman, the United States, 1.9.

Richard Horton. "This was an extremely hard editorial choice." To Murray's relief, Horton had decided that he and Lopez made the cut. Scientific studies based on burden of disease took off. Together the first four *Lancet* papers on the global burden of disease, published in 1997, would be cited more than 13,000 times.

Citations are more than just professional chits. They are a sign of influence, the way a later author justifies new research by acknowledging its debt to prior work. Lots of citations mean you are in some way directing the conversation, shifting the course of scientific investigation and the action that comes out of it. The outsiders with their radical new approach to measuring world health were becoming authorities.

But Murray's commitment to his work and colleagues didn't come without costs. He and his wife, Agnes, were increasingly estranged, by different interests as well as his long hours away from home. In late 1997, very soon after the birth of their third child, Amélie, they separated. A court battle followed over how and when Murray could see his children. Rival claims would cross continents as Agnes sold their house in Acton, moved with the children to the vacation house in Maine, and then eventually returned with them to France. By the time Agnes and Chris were officially divorced, she was in Clermont-Ferrand again. And he was in Geneva.

For almost fifteen years, Murray had circled the World Health Organization without quite joining it, sometimes a critic and sometimes a collaborator. Even after his separation from Agnes, it seemed unlikely that he would ever leave Harvard. In early 1998, the Department of Population and International Health at the School of Public Health made him, at the age of thirty-five, a full professor. He had gone from research fellow to the highest faculty level in ten years, during the same period attending medical school and completing an internship and residency. Grants he'd raised made his unit at the Pop Center financially self-supporting, staffed with six to eight Ph.D. students or fellows at any given time. Morale was high as the team prepared a multivolume book series on all their Global Burden research to date. First, however, they issued a fifty-page popular briefing meant for a much broader readership than the series. One of the 100,000 copies went to Gro Harlem Brundtland, who had served three terms as prime minister of Norway, most recently from 1990 to 1996.

By original profession a physician, Brundtland was the rarest of politicians: one who believed in independent science almost as much as Murray and Lopez. She had been the first woman and, at age forty-one, the youngest person ever to hold her country's office of prime minister. Norwegians knew her simply as Gro, or, more warmly, "Landsmoderen," meaning "mother of the nation." "Too often—in too many places—public attention is captured by the most vocal advocates," she would say after reading about the Global Burden of Disease project. "Large numbers of people suffering from serious problems are often ignored." When she said this, in 1998, Brundtland had just been elected director-general of the WHO.

The World Health Organization was, by this point, a very troubled institution—"unfocused, even corrupt, and overrun by middle-level management," people told Brundtland.* Yet it was also arguably the most important specialized agency of the United Nations, one with enormous sway over health policies worldwide and with the proven potential to improve the lives of billions.

---

*Accusations of petty corruption at the World Health Organization had become public in 1993. An external audit found that more than half of the voters for the winning candidate for director-general in that year's election had received "research contracts" paying as much as $150,000 with little work required.

Between 1967 and 1977, most famously, well-coordinated WHO vaccination and containment campaigns had led to the successful global eradication of smallpox, a disease responsible for as many as 500 million deaths in the twentieth century. And in 1978, WHO leadership had issued a landmark call for "health for all by the year 2000," presenting a vision where all human beings, regardless of wealth, would enjoy the highest possible level of health.

Brundtland invited Murray to Washington, D.C., for a cocktail party at the Norwegian Embassy celebrating her election. She had a mandate for a large-scale organizational reform. In the 1980s and '90s, the movement for "health for all" had stalled. The WHO had lost influence to UNICEF and the World Bank, and become known for a sleepy—some said close-minded and complacent—culture. It was telling, for example, that in the response to the greatest infectious disease threat of a generation—the global AIDS pandemic—the WHO had been largely sidelined and a whole new international health institution, the Joint United Nations Programme on HIV/AIDS (UNAIDS), had been created. To restore its international credibility and authority, Brundtland would write, the WHO required "a small revolution."

The day after the cocktail party, Brundtland met Murray one-on-one at a restaurant, the Jolly Roger, across the street from the Watergate Hotel. She told him things he already believed. "You manage what you measure. You only improve when you have numbers." It was a recruitment speech that, freely translated, went, *Tell me what the WHO should do. And then join me to do it.*

"That," remembers Murray, "was a very appealing challenge."

With Julio Frenk, Murray proposed a new top-level WHO cluster called Evidence and Information for Policy. "As health became an ever bigger part of the world economy, it was much more visible politically," Frenk recalls. One of the major roles the WHO could play, he and Murray said, was producing good evidence for policy-making at a global level. Succeed, and Geneva would become *the* source for accurate, up-to-date information and advice on all things health: what killed people, and what could save their lives; what made people sick, and what could make them better; how much we spent on health, and how we could spend it better.

Brundtland embraced it all. Murray, approvingly, watched her reshape the WHO's bureaucracy. Programs to address mental health and leading noncommunicable diseases—cancer, cardiovascular diseases, and tobacco-related illnesses—were given equal ranking with those for communicable diseases such as malaria and tuberculosis. And Brundtland started new programs to improve the quality and assessment of national health services, to strengthen the WHO's own technological capacity, and, last, to form the Evidence and Information for Policy cluster almost exactly as Murray and Frenk had proposed. "Her job is to set policy and strategy," aides to Brundtland told Murray. "Your job is to deliver."

Murray asked Harvard for two years' leave. As he packed up at the Pop Center, staff saw him carrying business books on leadership.

CHAPTER EIGHT

Taking on the World

Push and pushback—Going bigger—"A spy plan."

On July 21, 1998, Chris Murray became a director in the new WHO Evidence and Information for Policy cluster, heading a professional and administrative staff of approximately 150 people. One unit, led by Alan Lopez, would measure health loss and double-check other groups' statistics, using an expanded version of the Pop Center burden-of-disease team. A second unit would track health spending. A third, with a cost-effectiveness focus, would say what investments returned the most well-being per dollar spent. Another 100 to 150 people, outside Murray's direct oversight, conducted related investigations in health systems and health research, or managed the WHO library and publications. Above all of them, including Murray, and reporting directly to Gro Harlem Brundtland was Julio Frenk, from Mexico.

Actually, Frenk's job, leading the entire cluster, had been intended for Murray and was his to lose, which he accomplished with his usual speed and efficiency. Immediately on arriving in Geneva, he had sparred at a dinner party with a major anti-tobacco advocate. "I was young and aggressive," he would remember later. "People I perceived as chronic bullshitters irked me." His response, "which was not always appropriate," he admitted, "was to seek to point out they were wrong." Word got back to the boss. Brundtland did not like drama. "The way she likes to run a government is to get a team of young powerful horses," her aide-de-camp told Murray. "They're trying to gallop off in all directions. She reins them in." "I was ruled out in favor of Julio," Murray says. "It was the right decision at the time."

Even becoming a "mere" director was a massive transition for Murray. One day, he had been a young faculty member overseeing six research fellows in the almost invisible annex of an independent academic research center in Cambridge, Massachusetts. The next, he was a highly visible and influential figure in the most powerful health organization in the world. If you were a professor, you might pass your entire career without being able to influence national ministers of health. Now they were close enough to reach by whisper—if, that is, they would listen.

In bureaucracies, it is said, nothing is allowed, but there's a way around every rule. Murray, the skeptic and debater, now had to master the more subtle art of office politics. Anything sensitive, for example, had to be said in person, not in writing, or it might be used against you. Want to give a contract to run a meeting in South America? There was the official procedure that took four forms and six weeks or the critical phone call that reduced this to a couple of hours. Early on, a staffer came to Murray's office. He wanted Post-it notes. There was a rule, though, that certain supplies were allowed only for directors and above. "Post-it notes had to come from me or from Boston," Murray said. Every document that required a signature at the WHO came in what was called a "signing folder." The director-general used purple or green ink; executive directors had to use some other color; unit chiefs had a third color, and mere mortals a fourth. Murray took the first signing folder he received out to his assistant and said, "I don't ever want to see one of these again." The next day, the stack was back, only taller. "I lost, they won," Murray remembers. "That was a rude awakening to reality."

The arrival of Murray and Frenk was also a big change for Geneva. The World Health Organization had never had anything like a policy unit, only separate programs specializing in the fight against different diseases. Each disease had its own experts and its own estimation methods. Within that narrow sphere, policy recommendations were specialized, not holistic, based on data sent in by member governments, data that might or might not be reliable. On day one, Murray announced that the WHO, for the first time, would make its own official estimates and projections of illness, injury, and death. He wanted not only to replicate but surpass the advances he and his colleagues had made for the World Bank and at Harvard. "We have to do completely new estimates," he said. "We have to develop completely new methods." The WHO would be independent, impartial, innovative, and comprehensive; it would supply information others lacked or had wrong.

In other words, the WHO would now be not just a policy organization but also a center for sophisticated data analysis. Everyone who worked for Murray was clustered with him in space on the third floor of WHO headquarters divided into small bays by movable white modular walls. The technology at hand was also no better than ordinary office standards—"pretty crappy" desktop computers, Murray would recall later—but he wanted his staff to replace rosy health summaries from member governments and the unvetted statistics of nongovernmental organizations, outside UN agencies, and individual WHO programs with figures drawn directly from national vital registration records, hospital files, household surveys, demographic assessments, and finance reports. As long as people were suffering and dying, the Evidence and Information for Policy cluster would be like an international 911 call center, identifying distress and tracking the response. If governments and aid groups could not—or would not—supply honest estimates, the WHO would do it for them, and show them to the world.

The problem for anyone working with Murray was that, as ever, this meant late hours, weekend hours, and holiday hours. "That was how Chris expected people to work because that was how he worked," Alan Lopez says. He and Murray had, by this stage, collaborated for fifteen years, and no one subscribed more enthusiastically to the new program. "That someone of his genius would come to the organization and try to change the practice and output and reputation of the WHO was fabulous," Lopez thought. Nonetheless, "It put a lot of stress on me," he remembers. "I had to manage people's performance upwards in the organization, so everyone was contributing, and, at the same time, it was very hard to get them all to perform at this level."

WHO staff were international civil servants, used to working fixed hours, approximately 8:30 a.m. to 5:00 p.m., five days a week. "They did things according to a conventional schedule, both in terms of time and output," says Lopez. "They spent a lot of time in meetings." The kind of competitive drive common in academic and corporate science was alien to the office. Many of the employees Murray had inherited from the previous health statistics and health systems research divisions reacted to his directives with disbelief. "They came and went at the same hours as they had," says Lopez, "hoping that Chris would move on and the WHO would return to the nice place it was."

The human animal can accomplish amazing things under emergency conditions, but few people want to live—or work—in a state of perpetual crisis. As his new colleagues and subordinates discovered, however, Chris Murray was one of those few who relished the frenzied pace. In part, he was an adrenaline junkie gifted with extraordinary vision, ambition, and endurance—both physical and mental. Neither a medical residency nor a double black diamond ski run exhausted him. In part, too, though, he was compelled by a sense of the relative insignificance of one individual's personal life measured against the urgent needs of the rest of humanity. His obsession with work, perhaps compounded by the extreme pressure he was under to deliver on what he had said he would do, had already contributed to the end of his marriage and cost him contact with his children. He was not about to be about cowed by the workplace customs of Geneva.

Instead of giving up in frustration, Murray responded to the leisurely traditions of the WHO by creating a shadow staff of temporary workers who would embrace his sense of urgency and, he hoped, energize the larger institution. Advertising in universities worldwide, he offered short-term appointments to other work addicts from a broad range of fields who were willing to accept one- to three-year contracts to come to Geneva and help "provide an objective assessment of the various types of evidence which should influence health policy." The positions were funded by the UN Foundation and the Rockefeller Foundation.

Almost one thousand people worldwide applied to be "global health leadership fellows." Twenty-four were handpicked by Murray personally to join his team. Though a few would make careers at the World Health Organization, none started as official staffers. Instead, as the job description put it, "Fellows will not be employees of WHO but will be attached to the Organization for a limited duration."

As a member of Murray's statistical SWAT team, you might arrive at 9 or 10 a.m. This was a little after the regular staffers. But you left at 7 p.m., 10 p.m., or midnight—if at all. Esprit de corps among Murray's high-level temps was built on working night and day on specific projects before moving on to another situation, never necessarily integrating with WHO lifers.

Josh Salomon, for example, who had worked in Murray's

burden-of-disease unit at the Harvard Pop Center, had already entered a graduate program in health policy and decision science when Murray recruited him with the unique arrangement of a posting in Geneva ten days out of every month. His first assignment was a comprehensive review of international statistics on HIV/AIDS. "When I came, I'd work as much as I could," Salomon remembers. He had plenty of company. On the eighth floor of WHO headquarters was a break room near a shower. "Some of the young kids who came here working for Chris *lived* there," Salomon says. "That was their place of residence." One guy, he recalls, bragged that he hadn't left the building for a month. The key to the fellowship program, the way it got so much work out of people, was that everyone understood it was temporary.

Some of the career professionals in Evidence and Information for Policy embraced the new pace and tone. Yet resentment among those who safely enjoyed the benefits of a WHO contract— "home leaves, nine-to-five hours, comfortable ski weekends, laid back, well paid," Lopez summarizes—ran high. Assignments they had once had years to complete were now expected in weeks or months—or given to someone else. The evidence for every finding would be vigorously scrutinized by Murray and his new whiz kids. "He was brash and bold," says Lopez. "He changed the nature of the organization. He made it more scientific. He made it more accountable. He pushed people to excel. He pushed the WHO to raise its standards."

The push—and pushback—hardly ended in the offices of Evidence and Information for Policy. In creating the new department, Gro Harlem Brundtland had had to take money from other WHO clusters. Many in Geneva therefore viewed Evidence and Information for Policy as a threat. Technical programs could no longer make public claims about death or disease tolls without a check to be sure that they were logically defended and internally consistent. That external verification fell to Murray's team. When Josh Salomon walked into meetings, he heard muttering: "Oh, the thought police are here," people said.

Outside Geneva, other UN agencies were equally upset. They didn't want the WHO to have the last word on statistics that determined aid distribution, national health priorities, and lives saved or lost. "By 1999, between the UN Population Division, the UN Statistics Division, UNICEF, and the WHO, you had four different estimates of child mortality in Zambia," Alan Lopez recalls. "There was massive tension." Improving the health prospects for children required a common understanding of where they already stood. Were deaths increasing or decreasing in Zambia, for example, and at what rate? "When there's confusion about that, it makes it much harder to garner the political and social consensus to take action," Murray acknowledged. But the solution to confusion was not to have everybody agree on the wrong number. "That will reduce confusion but have people make the wrong decisions," he said. "You can't understand who's doing a good job and who's not." As long as Brundtland and Frenk backed him, Murray refused to give up any part of health measurement.

In fact, he wanted to go bigger.

By early 1999, six months into his appointment in Geneva, Murray had decided measuring health alone was not enough. To really improve lives, he thought the WHO should lead the world in larger health systems analyses like those he'd contributed to at the Harvard Pop Center and for the World Bank. Even the latest Global Burden of Disease study, for all its seeming comprehensiveness, told only about the health *problems* people faced, and only by large global regions. How societies tried to solve those problems was through health systems—specific research centers and public health institutes, hospitals and clinics, home care and traditional healers. If you were sick or injured as a child in Zambia, a young adult in Argentina, a head of household in India, or an elder in Italy, statistics didn't matter to you—you and your family just wanted help getting better. Murray repeated his question: *Who's doing a good job and who's not?*

His chosen vehicle to answer that question, with Brundtland and Frenk, was the next edition of the WHO's annual *World Health Report*. The 1993 *World Development Report*, initiated by the World Bank and then co-sponsored by the WHO, had introduced a new single measure of international health; the 2000 *World Health Report* would put forth a single measure of national health systems' performance. The WHO would rank its own member nations' achievements, best to worst, on a numerical scale.

The idea was wildly ambitious even by Murray's standards. Working with Frenk, he defined national health system performance not just by how healthy or unhealthy a country's people were, but also by how much health was improving or declining, how small or large the gaps were between the best and worst off in a country, and how well services matched demand. If you spent less money, public or private, to achieve the same results as other nations, you would rise in the rankings. If you spent more, you would fall. Rankings also dropped for health systems that drove the sick to bankruptcy or excluded the poor from care at all.

Before Brundtland took office, the WHO had issued a statement to *The Lancet* that Murray and Lopez's views "do not reflect the opinions, policies, or standards of WHO." DALYs and the burden-of-disease approach, it said, "are potentially useful for health situation assessment but require further research." Now, when insiders found out Murray's team planned an international ranking with an all-new WHO-produced Global Burden of Disease study as just one of several complex components, an even more intense counterblast began.

"The bureaucrats went ballistic," Murray would recall later. "People were pretty vocal: 'It will create waves with member states. It will create controversy.' There was both open conflict and the bureaucratic strategy where you try to block things." Support from the very top, though, never wavered. Brundtland stood with Murray and Frenk. Their rankings, she believed, would show off the revamped organization's new scientific authority. "She quite bravely said, 'No, no, we should proceed,'" he remembers. What Brundtland hoped was that the new information would trigger massive changes—improvements—in health systems around the world. That, after all, was the underlying goal of the WHO and every other international health organization.

Newly encouraged, Murray now demanded even more from both trusted members of his permanent staff and the short-term fellows. Every day came a fresh push for bigger and better data collection, and faster and more reliable estimates and analyses. In the fall of 1999, when Murray had been at the WHO for fifteen months, the members of his working groups were going home regularly at 7 p.m. Come winter, that was pushed to 10 p.m. After the new year, competition was fierce for middle-of-the-night vending machine fare and everyone on the project, not just the specially recruited global health fellows, was keeping a toothbrush at the office. The WHO, on the outskirts of Geneva, was surrounded by farmland. Across the street, where the new headquarters of UNAIDS would soon be built, was pasture. At dawn, members of the team would hear the bleats and crows of waking goats and chickens.

Had Murray ever read those management books he brought to Geneva? Even the most dedicated, idealistic fellows may have wondered if they were being led by a madman. On Saturday and

Sunday, with most lights at headquarters turned off and most offices empty, Murray manned a Razor scooter, 2000's fad holiday gift, on which he careened up and down the third floor's long halls, checking in on people. As much as he was becoming a seasoned project director, he was still the kid who'd told his college roommate to commute to campus via unicycle, and then took him downhill skiing for the first time in order to drum up fourth-class ship passage to North Africa. If you followed Murray's example, you would conclude that the only things worth doing were adventures. And it wasn't a real adventure until you'd reached your physical and intellectual limits.

Murray visited each working group independently, "almost like a spy plan," Alan Lopez remembers. "He had cells of people working together, but independently of other cells. I knew what I was doing, but not what the health economists were doing, and vice versa." Murray's purpose may have been keeping each cell single-mindedly focused on its own tasks. But working in the dark is not a good way to maintain morale. "We're getting more pressure from Chris to get the methods and measurements right," Lopez recalls. "It's very tough business. Things are getting grim. People are working later and later hours." Murray, he says, was "just driving people to despair."

Finally, one night in February, seeking to rally his team, Murray gathered everyone in one large room. It was midnight, Lopez remembers. "He explained to us for the first time how it all fit together. How he planned to integrate all the pieces we were working on and why they mattered." For the first time, people understood their group's role in the larger project. They began to see what Murray saw in their statistics: parents hoping their children survived a deadly illness, young adults faced with a serious accident and injury, a grandfather susceptible to stroke at the same time his caregiver suffered back pain and depression.

They also had hundreds of questions, however, about how each component of the international rankings would be determined. Weighting individual factors such as "health system responsiveness" and "fairness of financial contribution" was totally new. For better or for worse, wouldn't the 2000 *World Health Report* would be the WHO's version of *U.S. News & World Report*'s college rankings or *Motor Trend*'s "Car of the Year" issue: impossible to ignore, yes, but certain to be disputed?

"It was a lesson to Chris," says Lopez. Even the people closest to him weren't convinced he could win this argument. "He began to understand that while this was a phenomenal achievement in terms of measuring population health and health systems performance, it was going to be a very, very controversial and difficult thing to sell." And if the people on his team had questions, what about the rest of the world? Statistics didn't just have people behind them—they had people in front of them. Political leaders. And these were the people who ultimately oversaw the WHO.

CHAPTER NINE

No One's Sick in North Korea

Front-page story—"Persistence beats resistance"—A fly on the wall—The cubbyhole.

The 2000 *World Health Report* was released on June 21, 2000. It would take the WHO's new evidence-based approach to health policy to a wider audience and was written with that in mind. The first 140 pages were presented in even-handed, high school textbook-style prose, supported by figures, and organized as answers to basic questions: "Why do health systems matter?" "How well do health systems perform?" "Who pays for health systems?" "How is the public interest protected?" The next 60 pages were statistical tables evidencing rather astonishing research for anyone with an eye for numbers. At the end, official "Annex Table 10," page 200, came rankings of overall health system performance for individual nations. And that, of course, was all most people read.

The table of relative rankings made global headlines and is still cited by journalists, politicians, policy analysts, and editorial writers around the world. Ranked in first place was France. Four other wealthy Western European nations followed. Singapore ranked 6th. There were at least two surprising stories. The

Middle Eastern sultanate of Oman came in 8th, beating Austria, Brundtland's own Norway, and the nation with the world's longest average life spans, Japan. Colombia was 22nd, a notch above Sweden and three above Germany.

With rankings, however, every winner meant a corresponding loser. The United States ranked 37th overall, between Costa Rica and Slovenia. (It had been first in health system responsiveness, but 24th in healthy life expectancy—equivalent to how long an average newborn, faced with today's standards for sickness at every age, could expect to live in perfect health—and 54th for fairness of financial contribution, a measure of how many households could not afford health care or risked impoverishment from its costs). Two emerging powers, India and Brazil, ranked 112th and 125th out of 191 nations. Russia was 130th. China, a public health darling for at least the previous two decades, was 144th, behind even Haiti.

The media loves lists. Coverage of the 2000 *World Health Report* was fast, furious, and international. "U.S. Spends More Than All Others, but Ranks 37 Among 191 Countries," headlined *The New York Times*. In Malaysia, rated 49th, "WHO's ranking 'not accurate,'" said the *New Straits Times* of Kuala Lumpur, quoting the Malaysian deputy director-general of health. A *Wall Street Journal Europe* commentary compared Murray with Karl Marx. Critics on the left in Brazil, meanwhile, said he was part of a larger conservative political project obsessed with "reducing the size of the public sector, increasing the participation of the private sector, privatizing and delegating decision-making to independent agencies."

Murray, Frenk, and Brundtland welcomed the debate. Arguments meant productive attention. Governments turned over all the time because their economies rose or fell. Let politicians compete on their health record, too, they said. Of the United States,

Murray was quoted as saying up to 10 percent of Americans had a life expectancy of fifty years or less; the French recorded much better outcomes at almost half the cost. Oman's leap was based on "a well-planned upgrading of its health care facilities and services"; China's drop was because high out-of-pocket payments now kept most people from getting effective care. Colombia won the top spot among Latin American nations because a graduated health insurance scheme meant coverage for citizens able to pitch in as little as a dollar each per year.

In terms of guiding health system improvement for all, the 2000 *World Health Report* was explicitly a beginning, not an end. "The material in this report cannot provide definitive answers to every question about health systems performance," Brundtland wrote. "It does though bring together the best available evidence to date." Decades of clinical trials helped doctors give consistent, evidence-based advice and care to their patients. No such tradition existed for comparing and improving the performance of larger health systems. To those who didn't like the new rankings, she told attendees at a London press conference, "We say, 'Come and help us refine and improve the analysis next year, and the year after.'"

The WHO, however, was not a Pop Center seminar. Its decision-making body, the World Health Assembly, was made up of 191 national delegates—usually ministers of health—each responsible to his or her own head of state, with an average tenure in their jobs of less than two years. Embarrass the leader at home and that time might be even shorter. José Serra, the Brazilian minister of health, was running to be his country's president. "Because we said Brazil wasn't performing very well, he took offense," Murray remembers. "There was a long process with Brazilians coming to the WHO, saying to the Secretariat, 'You have no right to do this work.'"

"The idea behind the 2000 *World Health Report* was a radical one," Josh Salomon would say later. "The WHO would somehow hold its governors—the member states—up to be accountable." He sighed. "It's just not the way the WHO works. In its governance structure, there is nothing beyond the member states." The same politicians who hated it when Murray gave their nations' health systems low rankings were his boss's boss's boss.

Serra, claiming his country's ranking was political sabotage, worked hard to get Frenk and Murray fired. Because they had elected Brundtland, other ministers also brought their objections to her directly. "That's the price of a policy unit," said Barry Bloom, Murray's former collaborator on tuberculosis research and a longtime adviser to the WHO. "Once you start pontificating, you have one hundred and ninety-one people who may not be happy about what you're saying." In the summer of 2000, Frenk left the WHO to become minister of health of Mexico. Despite the controversy and an appealing opportunity to lead the department of public health at Oxford University, Murray decided to accept Brundtland's offer to replace Frenk as executive director of the Evidence and Information for Policy cluster, reporting only to her.

For much of the next year, Brundtland sent high-level representatives from Evidence and Information for Policy around the world to explain themselves and answer concerns about the new health systems assessments. The WHO called this process "country consultation." Outcomes were mixed. Sometimes the conversations led to better data. Other times politicians screamed at you. "There were very, very diverse opinions, a lot of emotions, a lot of anger," Alan Lopez remembers. Wherever they went, people were split. "Some agreed with the methods," says Lopez. "Some said you could do it differently or better. Others said you shouldn't do it at all unless you have more data."

In a few countries—Pakistan, ranked 122nd, for instance—authorities protested almost every aspect of the 2000 *World Health Report*. Most governments, though, just wanted one or two sensitive estimates "tweaked." One particularly surreal instance was the North Korean delegation's complaints that estimates of their countrymen's healthy life expectancy had to be incorrect. "Healthy life expectancy in North Korea is the same as life expectancy, because nobody is sick," they said. Murray tried not to laugh. "What do you do?" he says. "Of course we paid no attention to it, but it was one of those very bizarre moments for the translator."

He was temporarily reining in his instincts to argue. "Listen, and before you speak, think about the pros and cons," Julio Frenk had advised him. "Persistence beats resistance." When agreement was impossible, however, the WHO returned to publishing two sets of numbers—the Indian government's take on childhood vaccination coverage, for example, or what Ethiopians said was their HIV infection rate, side by side with estimates from its own statisticians. That was "very instructive," Murray thought. "Countries care very much about certain diseases that have political salience. You either have to have strong leadership that's willing to withstand criticism, or be very cautious if you're at the WHO."

At Murray's suggestion, the WHO formed an independent scientific peer review group to review the entire process of assessing health system performance. Its chair was Sudhir Anand, the Oxford economist who'd debated Murray so fiercely at the Pop Center. "It was a very high-stress, intense period, battling to survive, to keep that analytic work alive," Murray would remember later. He compared the resulting bureaucratic and scientific back-and-forth to a chess game: "We knew where we were weakest and most vulnerable to criticism and we raced to fill our gaps."

In the midst of this, Richard Horton, the *Lancet* editor, came to Geneva for a week. It was the first time he and Murray had met in person, and relations were tense. Horton thought Brundt-land was an awkward, autocratic leader and considered the 2000 *World Health Report* a disaster for the WHO. "They'd upset all these countries," he would recall later. "It was the worst sort of paternalism and arrogance. I was there to be a fly on the wall and see how this had happened."

Horton, born in London in 1961, was Murray's age. He'd trained as a doctor at the University of Birmingham and joined *The Lancet* as an assistant editor in 1990. In 1993, he became the journal's North American editor, based in New York City, and in 1995 he moved back to England to take charge as editor in chief. Since then, what was already a premier outlet for research had become a scientific journal that also took a uniquely activist role. Horton sponsored investigative journalism and often led exposés. He published spirited editorial calls to action and opined freely about the imbecility of political, academic, and bureaucratic obstacles to progress. "He's become a power," Murray would say later. "American editors think it's an outrage, but if you want to influence global health generally you have to be in *The Lancet*." Go to a UN conference, he continued, and on the dais "they'll have two heads of state, the director-general, and Richard Horton."

The Lancet had already given space to the Brazilians and other critics of the 2000 *World Health Report*. Soon they would print an essay from the report's editor in chief, Philip Musgrove, based now at the U.S. National Institutes of Health, saying Murray and

Frenk had excluded him from key decision-making and disavowing the rankings altogether. Horton's own report, titled, "World Health Disorganization," said many staff were "embarrassed . . . to be associated with this highly criticized product." Horton was being Horton, "fostering public discourse," Murray would say later, but "there was a period where *The Lancet* was not our friend."

In May 2002, the scientific peer review group chaired by Sudhir Anand submitted its final report to the director-general. The findings would be crucial to Murray's reputation and future as a public health leader. Far from supporting all the critics, Anand and his colleagues endorsed Murray's work. "The objectives of the health systems performance assessment initiative are valid," they concluded, and "the provision of comparative data on health-system characteristics is a vital component of securing health-system improvements." The 2000 *World Health Report*, they said, made "an important breakthrough in seeking to provide an integrated quantitative assessment of health systems performance, and bringing the topic of health-system performance to the attention of policy makers worldwide."

It appeared Murray had faced down his first serious career crisis—and not only survived, but prevailed. And so had the idea of new indicators such as healthy life expectancy and health system equity and efficiency, though not without making sure that there would be some adjustments in the future. Going forward, any first-to-worst rankings would be abandoned, but the WHO said it would update Global Burden numbers annually and evaluate national health systems on an A-to-F letter-grade scale every other year. In a fitting irony, the entire controversy had become a way to get Geneva to hire an outside group of experts to tell Murray's team how they could improve. "Our plan had been: strengthen the empirical work, strengthen the methods," Murray would say later, claiming the entire consultation-and-review process as a victory for his cause. "We had."

If the object was to spur nations to improve their health systems by making them compete, they had certainly succeeded in the first step—getting their attention. Whether the World Health Organization would continue to support this maverick, with his outside funding, his shadow staff, and his barely suppressed conflicts with the politicians between him and the people he wanted to help was another question.

In August 2002, Gro Harlem Brundtland announced, to everyone's surprise, that she would leave the WHO the following July. She was sixty-three years old, she had served in high and demanding offices for three decades, and she wanted to spend more time with her family.

Murray and others who depended on her backing were devastated. In their eyes, Brundtland was the rare public health leader who understood the hidden costs of not paying attention to everything—the orphans produced by neglecting AIDS and tuberculosis, the adults saved by vaccines but killed by traffic injuries, the epidemics such as anxiety and osteoarthritis that went untreated because they were absent from mortality statistics, and the billions who never sought necessary care because they could not afford it. Her departure threatened the revolution in thinking she'd encouraged. The fate of policy recommendations based on actual conditions rather than on the most visible—but possibly not most burdensome—suffering was in doubt. "She should have and could have served another five years," Murray thought. "She'd come in and done this huge reform. You can't change an organization like the WHO in five years." And without Gro Harlem Brundtland remaining in charge, the whole idea of Evidence and Information for Policy might be scrapped.

As is true of almost any important UN post, candidates to be WHO director-general have to be put forward by their country. It was unthinkable, however, that anyone forwarded by the United States would be chosen as leader. America was already too powerful. Murray therefore asked the New Zealand delegation if they would nominate him as a representative of their nation. "Me being as much an American as a Kiwi, they said no," he recalls. His hard-driving leadership style and well-known friction with other member countries could also well have been factors in the refusal. In any case, Murray then joined others in lobbying his old boss, Julio Frenk, now the minister of health of Mexico, to be a candidate. Frenk agreed. "At that point, you're doing your job, watching what unfolds, and hoping like hell Julio is going to be elected," Murray says.

In the weeks before the January 2003 election, competition narrowed to three top candidates: Frenk, from Mexico; Belgian-born Peter Piot, the executive director of UNAIDS; and a long shot, Jong-wook "J.W." Lee, a twenty-year WHO veteran from South Korea who was in charge of tuberculosis programs in Geneva and among those public health leaders who favored focusing on a few major maladies rather than taking them all on at once. According to the election procedure, voting was conducted in rounds; one at a time, the candidate with the least votes was dropped from consideration. Frenk fell early. He and Murray, encamped together at the house of the Mexican ambassador in Geneva, where they could listen to the ongoing tally by phone, began rooting for Piot. "He believed in measurements and evidence," Murray thought, "and using numbers strategically whether countries like it or not." Two tied votes followed. Then someone switched sides, choosing Lee over Piot. "That was that," Murray says.

Lee would take office in July. Murray spent six months not knowing what was going to happen. Was the new reality as bad as he feared? "You never quite know," he said. In the entire transition period, neither Lee nor his staff ever said a word to Murray, good or bad. Would the WHO revert to being mainly an administrative arm of the assembled governments, accepting old priorities and received information, or would the push for independent evidence to ensure the greatest possible good for the well-being of humanity be continued? "Up to that last week, there was a kernel of hope that the new guy would buy into it," Murray remembers.

The day before Lee took office, the kernel was crushed. "Some minion from his office said, 'As of tomorrow, you're moved,'" Murray says. Lee kept the Evidence and Information for Policy cluster, but redirected core measurement efforts from Global Burden to the Millennium Development Goals, a recent set of international resolutions to reduce child and maternal mortality and reverse the spread of HIV/AIDS, malaria, and tuberculosis, among other objectives. Replacing Murray as director was Tim Evans, a Canadian who had been with him at Oxford as a Rhodes scholar and at the Pop Center as a fellow. "It was very awkward," Murray says. "Someone I'd known for years took over my job and we never talked about it."

Lee, Murray learned much later, had forbidden Evans from speaking to him. Adding to his isolation was the absence of Alan Lopez. Lopez had left Geneva in December 2002, after twenty-two years of service at the WHO. At the age of fifty, he had been recruited to be the dean of the newly established School of Population Health at the University of Queensland in Brisbane, in northeastern Australia—about as far from WHO headquarters as you could get.

The burden-of-disease team in Geneva was slashed from twenty-two staff members to two. Murray was shunted from his executive director's office to a little cubbyhole on the way to the cafeteria. "It was pretty brutal," he remembers. "I had nothing to do," no staff, no responsibilities.

He was forty years old. Murray, who liked nothing more than to work, to explore, to learn, and to lead change, was given the role of "adviser" with nobody to advise.

CHAPTER TEN

Racing Stripes

"Pay attention, pay attention, pay attention"—A bold proposal—52 million Mexicans—Switzerland and Somalia—A tale of two Larrys.

This was not ancient Rome, when a single power dominated the known world and a leader sent into exile might as well have fallen off the edge of that world. Geneva was the headquarters of the World Health Organization, but the WHO no longer monopolized discussion of life and death. By 2003, global health in general and the battle against AIDS in particular had become a cause célèbre and a celebrity cause. Where small groups of activists and organizers had led, billionaires, presidents, and prime ministers now followed.

In late 1999, Bill and Melinda Gates, the world's richest couple, had donated $750 million to launch the Global Alliance for Vaccines and Immunization, or GAVI, a public-private partnership headquartered in Geneva and Washington, D.C., whose mission was to deliver necessary vaccines to children in poor countries. Between 2001 and 2002, the Global Fund to Fight AIDS, Tuberculosis, and Malaria was formed with an initial war chest of more than $1 billion, $100 million of it from the Bill & Melinda Gates Foundation and almost all the rest from G8 governments (the United States, Japan, Germany, France, the United

Kingdom, Italy, Canada, and Russia) and the European Union. And on January 28, 2003, President George W. Bush announced the President's Emergency Plan for AIDS Relief (PEPFAR), an extraordinary five-year, $15 billion commitment to treat 2 million patients, prevent 7 million new infections, and provide care for 10 million people affected by the AIDS epidemic around the world. It was easily the largest pledge in history to fight a single disease internationally. In his own country, Bush would be remembered for tax cuts and the war in Iraq. But in Uganda, as *The New York Times* reported, people were "terrified that when Mr. Bush left office, 'the Bush fund,' as they call it, would go with him."

Jim Kim and Paul Farmer, Murray's friends from his residency at Brigham and Women's Hospital, describe the life-transforming effect of these huge new programs on activists and aid workers like themselves. "Until then, it really seemed that no one cared about global health," Kim remembers. "No one was attacking these major killers. It seemed like Paul and I were voices crying out in the wilderness, 'Pay attention, pay attention, pay attention.'" Now he and Farmer had money and attention from the rich and powerful and a new generation of enthusiastic young professionals begging to work with them. The nonprofit they had co-founded, Partners in Health, expanded beyond Haiti to South America and Russia, and served as the subject of Tracy Kidder's best-selling book *Mountains Beyond Mountains*, published in 2003. In the book's wake, enrollment in internationally minded public health programs swelled.

If the WHO didn't want Chris Murray, Harvard did. Barry Bloom, his former TB collaborator, was now dean of the School of Public Health. Larry Summers, Dean Jamison's old boss at the World Bank who had served as the U.S. secretary of the treasury under President Bill Clinton, had become the president of Harvard University in 2001. Fresh to the job, Summers had been looking for big ideas, and Bloom proposed a major new health initiative. Summers loved the concept. Within his first six months as Harvard president, he gave a speech arguing that there were two things people would remember best about the first half of the twenty-first century. One was the revolution in life sciences. The other was changes in the developing world. "What brings them together," Summers said, "is global health."

Still, the idea stalled from 2001 to 2003. "Nothing happened," remembers Bloom. "He wouldn't launch a program unless it had a leader." That changed in 2003. Summers already knew Murray from the 1993 *World Development Report*. Supporting it, he said later, was "one of the more important things I've been able to do." "Chris wasn't the most bureaucratically smooth guy," Summers acknowledged. But that wasn't necessarily bad. "If doing numbers simply confirms all the prejudices one has before, there's no reason to do numbers," he liked to say. "The reason to do numbers is that some things will be surprising."

The process of hiring Murray had to go through a university committee that included people whose work his findings questioned, but Summers, who had his own abrasive streak and was no stranger to argument, pushed successfully to hire Murray over these objections. "Chris was very controversial," he remembers. "He wasn't the king of working with other people. He did things his way." Summers's view wasn't that Murray's critics were necessarily wrong, but that it was better to let many flowers bloom. "Rather than not Chris, we're going to do Chris and other things," he said. If global health was truly the field of the future, he wanted Harvard to be the best.

In September 2003, Murray arrived in Cambridge for the third time in his academic career. He was now not only a full professor at Harvard Medical School, the Harvard School of Public Health, and the Faculty of Arts and Science, but also director of the new Harvard Initiative for Global Health, or HIGH.

Beyond its clever acronym, most of what would happen at HIGH was yet to be defined. But with Murray in charge, *how* things would happen was already clear. Catherine Michaud, his old Pop Center team member, observed that Murray was exactly the same as when he'd left. "Chris is Chris is Chris is Chris, wherever he is," she said. "He's very open to debate and very demanding for you to produce something that makes sense." What she had missed most without him, she decided, smiling wryly, was the punishment. "I was working much less and less hard when Chris was not here. The other projects were not as demanding." Murray was as happy to see her as she him. "It was very hard being tossed out of the WHO," he admitted. "I was just grateful to have something meaningful to do."

With him came a longtime colleague, Emmanuela Gakidou, an expert on the measurement of health inequalities within countries and peer groups. In the early 1990s, as a Harvard undergraduate from Athens, she had majored in biology and neuroscience but hated lab work, and so joined the Pop Center burden-of-disease unit as part of a semester-long independent study during her junior year of college. A few diseases had yet to be assigned. Gakidou was told she was responsible for chronic obstructive pulmonary disease, or COPD, which she had never heard of. "It kills over a million people in China a year," Murray told her. "Great," Gakidou said, with her own bravado. She stuck with the project through graduate school in international health economics and health policy, perfecting her English by editing page proofs on three thousand-page Global Burden of Disease books.

In Geneva, where Gakidou worked as a health economist for the WHO, she and Murray had started dating. "Emm is mentally, physically tough," he said. Another natural athlete, she gave as good as she got. Murray, the ex–college skier, had returned to the slopes in Switzerland. He didn't know if it was him or the equipment, but he was phenomenally better—"fearless," people said. Almost immediately he immersed himself in "big mountain" skiing, going off lifts into semiwilderness, maneuvering around glaciers and crevasses. If you aren't serious, friends told friends, don't ski with Murray, or at least don't try to keep up with him. "He almost killed my son," said a co-worker. Gakidou followed him anyway. "We came off a piece of heavy new snow," Murray remembers an early outing. "It's getting dark. This isn't terrain you want to be on." Gakidou struggled. "How do you take something so easy and make it so hard?" Murray shouted, passing her. The instigation, if that was his intention, worked. "I got so mad," says Gakidou. "I just wanted to kill him. I skied straight down the mountain."

Murray had met someone as intrepid as he was. Even their affection for each other was competitive, as, for example, when they fought to sacrifice to the other the last bite of a chocolate bar. "It's passive-aggressive," someone told Murray once, referring to their relationship. "No, it's aggressive-aggressive," he replied. With others from the WHO in Geneva, the couple began vacationing in North America, tackling the toughest ski slopes up and down the Rocky Mountains, from Crested Butte, Colorado, to central British Columbia, Canada. Gakidou shared Murray's endless energy. "There are people who are very happy to go to the beach for a week and just sit there," she put it. "That's my idea of a nightmare." Murray could not have agreed more.

As one of the first academic global health programs in the country, HIGH was another case of jumping off into an uncharted landscape, creating a trail where none had been. The entire field—once *international* health, concerning just a few lead nations; then *world* health, the exclusive domain of inter-governmental organizations and a few big private foundations; now *global* health, by and for everyone—was new. Celebrities like Bono and activist organizations like ACT UP spread awareness. Students clamored for new courses. Grantors demanded new research. Murray expanded offerings across the university system and recruited and linked interested faculty from public health and public policy, medicine and philosophy, demography, government, and economics, but he had no money to continue the advances he'd made at the WHO: measuring the global burden of disease and beyond.

To build his health measurement work up again, Murray brought Josh Salomon and Catherine Michaud back onto his team, alongside new recruits. Gakidou opened new areas of research on the impact of specific health programs. Scattered contracts started to come in. The National Institute of Alcohol Abuse and Alcoholism gave the scientists a quarter of a million dollars to measure the burden of alcoholism and alcohol abuse in the United States. The National Institute on Aging gave them $7.3 million to conduct a worldwide burden study of older adults. The World Bank funded a new, cross-national road traffic injuries database. Julio Frenk, in Mexico, hired HIGH to evaluate health system reforms he'd pushed for since becoming the national minister of health. They were good projects, but too piecemeal for Murray's taste. "It wasn't the Global Burden of Disease," says Gakidou. "There wasn't this idea that we'd have core funds and every year produce these metrics." Without the mandate that comes with committed funding, she explains, "You can't invest in generating mortality numbers every year for every country because that's not what you're being paid to do."

But without reliable numbers, who knew what difference aid groups and governments around the world were making, much less how to do better? It was the question Murray had been asking for twenty years, with ever-larger populations—and budgets—at stake. In 1990, total development assistance for health had been $5.8 billion; in 2000, it was $10.9 billion; by 2010, it would be $29.4 billion. In the United States, domestic health care spending topped $1.7 *trillion*—almost 16 percent of the national economy; worldwide, the domestic average was more than 10 percent. How well that money was spent determined who lived and who died. Yet we tracked daytime soap opera viewership or online shoe advertising better than we did national or international health outcomes.

Murray envisioned Harvard as a permanent home for impartial, scientifically derived statistics, free from the constraints of the World Health Organization and other UN agencies. Time and again, these agencies had proven they could not monitor and evaluate their own member nations impartially and without interference. To take a single example, in reporting on the prevalence of malaria, the WHO had gone back to simply repeating countries' own reports. As a consequence, it said Nigeria had a rate of 30 cases of malaria per 100,000 people per year. But every year, more than 150 Nigerians per 100,000 *died* of malaria. Based on the death rate, the prevalence rate must be more like *30,000* per 100,000 people. In a 2004 article in the *British Medical Journal*, Murray, Alan Lopez, and Suwit Wibulpolprasert, a member of the WHO executive board from Thailand, insisted that the "WHO is ill suited for the role of global monitoring and evaluation of health."

What decision makers really needed, Murray, Lopez, and Wibulpolprasert suggested, was an independent global expert team, beholden to nothing but the truth. Without directly asking for the job of forming that team, they went so far as to suggest its budget: $50–70 million a year.

At the same time Murray was making the case that much more information needed to be gathered and interpreted if new life-saving efforts were to flourish, burden-of-disease studies on the national level were proving it. In 2000, when Julio Frenk took office as the minister of health of Mexico, approximately 50 million people in Mexico were without health insurance, about the same as in the United States. Another similarity between the two countries was that trying to pay for medical care was the leading cause of personal bankruptcy. "The problem was that access to insurance was a benefit of employment," says Frenk. "Anyone who was self-employed, unemployed, or out of the labor market"—half the population—"was on their own." Rural Mexicans lacked access to care. Even where the health system was present, even for those who were insured, it was under-funded, Mexico's 1990s national burden-of-disease analyses had showed; and it was entirely structured toward acute communicable diseases, which was the burden the country had had half a century before.

In response, Frenk proposed a new national insurance plan, Seguro Popular, or Popular Insurance. "For what was given priority, and in what order, we used the national burden of disease," Frenk says. "You want to cover those interactions that give you the highest gain." Medications for breast and cervical cancer, osteoarthritis and arthritis were covered, for instance. Where emergency care following a car accident was once an out-of-pocket expense, now it, too, would be covered. So would treatment for mental illness, childhood cancers, and cataracts, the leading cause of adult-onset blindness. Additional programs targeted the specific needs of women, HIV/AIDS patients, poor families, and rural Mexicans.

So grand a system realignment required additional investment—not only in staffing, facilities, medicine, and equipment, but in professional training centers, health-information-gathering capacity, public education, and regulatory protection. What Frenk suggested would more than double the budget of the Ministry of Health between 2000 and 2010. But the drag on the national economy from doing nothing was greater than the extra expense of enlarging Mexico's health system, and increased productivity would improve tax revenues. And solid evidence that health care costs bankrupted so many individuals and businesses created an outcry from the media and public. The 2000 *World Health Report*, overseen by Frenk and Murray, had ranked the Mexican health system 144th in the world in terms of "fairness of financial contribution."

In late 2003, the plan for Seguro Popular was passed by a large majority of Congress. Between 2004 and 2010, physicians per person would increase more than 50 percent in Mexico. In almost the same period, the availability of nurses in the country would leap 29 percent, rates of breast cancer patients completing treatment would rise dramatically, and the number of households forced into poverty by medical expenses would drop to less than 1 percent.

Mexican presidents serve a single six-year term, yet the reforms would continue under a new president and health minister, expanding even amid the 2008–2009 global economic crisis. "It's been a huge success," Frenk believed. By 2010, his country's child mortality rate would be fewer than 17 per 1,000 live births, almost half what it had been in 2000 and *one tenth* the figure in 1950. Finally, in March 2012, Mexico would achieve universal health coverage. "Now in Mexico there are fifty-two million people who were uninsured and who are now insured," says Frenk. "They now have coverage for diseases that we would probably not have included had we not had the evidence of the burden of disease."

Other countries began to use burden-of-disease studies to shape their health programs and show how best to allocate their resources. By 2006, more than three dozen other local burden studies had been completed, with increasing sophistication. If the original Global Burden of Disease study was analogous to the first map of the world, local burden studies were like GPS for public health. When the top-level analysis reached policy makers, the impact could be immediate, no matter what the political or medical systems of the country.

In Iran, injuries—led by road traffic accidents—turned out to be the leading preventable cause of health loss. The national transport minister was fired; new roads were built; police were completely retrained. Second-worst were mental health problems, and the study uncovered a horrific hidden epidemic of female suicides by burning, which was countered once it was recognized. Third was cardiovascular disease, and the government changed subsidies to supply households with unsaturated rather than saturated cooking oil.

In Australia, thanks to a national burden analysis, short-term therapy for depression became free. At the same time, universal prostate cancer screenings were cut. The widely promoted tests led to a large proportion of false-positive results, causing more harm than good as individuals were made to undergo painful, expensive, and unnecessary follow-up diagnostic procedures and treatments. In Thailand, a five-year burden assessment showed many more HIV deaths than officially reported and a rising tide of stroke and heart disease. Among many responses, the country would soon outperform most others in providing access to antiretroviral treatment, with a very large drop in HIV deaths, and its national insurance program added drugs to lower blood pressure and cholesterol. In nearby Vietnam, meanwhile, the government mandated motorcycle helmets, literally overnight, after seeing that motorized two-wheel traffic collisions led to a greater national burden than lung cancer, preterm birth complications, or tuberculosis.

Any thorough and sensitive burden of disease and cost-effectiveness analysis also weighed other considerations, such as what populations you wanted to reach most or who had first priority for treatment. Aware of their different population groups, Australian researchers completed a national study and, simultaneously, a separate burden study for the country's indigenous population, Aboriginal and Torres Strait Islander people. The gap in outcomes was enormous. Indigenous Australians suffered health loss from cancers at 1.7 to 1.9 times the national rate. The toll from suicide, violence, and unintentional injuries was 2.4 to 5.3 times as high. The impact of heart disease and diabetes was 4.4 to 6 times greater. If current rates continued, one in three Aboriginal and Torres Strait Islander teenagers would die before age sixty, the study said. For Australians as a whole, the rate was lower than one in twelve. Following the study's publication, the government announced a new emphasis on Aboriginal and Torres Islander health, from raising birth weights to curbing diabetes. From 2009 to 2014, close to $900 million would be committed to programs to reduce the Australian burden of disease nationwide. "Tobacco use has gone through the floor," says Jane Halton, secretary of health for Australia from 2002 to 2014. "Childhood obesity numbers are flat. Diabetes numbers are flattening."

Public health programs are always political and it is an uncomfortable fact that autocratic governments can sometimes make change faster than democracies. But Mexico and Australia were both democracies, and their health systems were the two where burden-of-disease analyses took deepest root. Leaders in both nations believed the investment would actually save them money. At the same time, in the United States, a history of resistance to any sort of unified health system brought with it economic costs to the whole country, not just the sufferers. "Evidence that other countries perform better than the United States in ensuring the health of their populations is a sure prod to the reformist impulse," Murray and Frenk would write in a 2010 article in *The New England Journal of Medicine*. "It is hard to ignore that in 2006, the United States was number 1 in terms of health care spending per capita but ranked 39th for infant mortality, 43rd for adult female mortality, 42nd for adult male mortality, and 36th for life expectancy. . . . Comparisons also reveal that the United States is falling farther behind each year."

One reason for America's falling standings was that health differences between different populations were even more extreme than in Australia. According to an analysis Murray led at Harvard, life expectancy circa the year 2000 for Asian American females in Bergen County, New Jersey, was ninety-one years. For Native American men in Bennett County, South Dakota, it was fifty-eight years. In the same period, there was a smaller gap between Switzerland and Somalia.

Forget the old distinction between "developing" and "developed" nations insofar as the burden of disease was concerned. All Americans could be said to live in a developed country. But with very different burdens. "Ten million Americans with the best health have achieved one of the highest life expectancies on record"—three years better than Japan for females and four years better than Iceland for males, Murray and his co-authors noted. "At the same time, tens of millions of Americans are experiencing levels of health that are more typical of middle-income or low-income developing countries." Wealth alone did not explain differences, nor did where people lived, or a single cause of death such as homicide or HIV.

Data, Murray was demonstrating again, had the power to ferret out important stories authorities would otherwise miss. He and his co-authors on the study concluded with another call for better health systems monitoring and reporting. In the United States as worldwide, they wrote, "It is when the public, community and professional groups, media, and politicians focus on what is being achieved, and why efforts are working in some places and not others, that the culture of accountability for health outcomes will be strengthened."

Murray's group at Harvard and another Alan Lopez had started at the University of Queensland completed or advised on most of these new burden-of-disease or local life-expectancy studies, but they were still far from having the fully funded institute that could take on the challenge of compiling data for everyone, everywhere, including places that might never contract for a study on their own. Yet even if $50–70 million a year was not an unreasonable budget for a huge global health monitoring project of the scope Murray and his associates envisioned, it was a good bit more than Harvard or any government grant was likely to provide. Could anyone in a new generation of science-minded multibillionaires help?

Through a HIGH donor, Murray met Larry Ellison, founder and CEO of the database software giant Oracle, in the spring of 2004. Then the twelfth-richest person in the world, according to *Forbes* magazine, Ellison had taken his company public in 1986, just one day apart from Bill Gates, his long-time rival, and Microsoft. When Murray gave his pitch for an independent academic institute to monitor and evaluate health programs, Ellison was enthusiastic. "He liked the idea of being the guy who'd fund critical analysis of the numbers in global health," Murray would remember later.

Ellison himself was a very successful loose cannon: brilliant, powerful, fiercely independent, charged with energy, and unpredictable to an extreme. Fifty-nine years old and more than six feet tall, sporting his graying beard and mustache trimmed to suggest a certain Mephistophelean raffishness, the Oracle CEO, on discovering Murray's love of outdoor sports, invited him to join the crew on his racing yacht during a weeklong America's Cup Class competition. This was fast company indeed. When Ellison's yacht won, Murray, their eighteenth man, found himself close enough behind his host to receive almost full-bore the first blast of celebratory champagne. It looked like a great triumph for Harvard as well. Afterward, Ellison asked Murray to put in writing a proposal for what he wanted.

A major job of the president of a university is to court donors. In the spring of 2005, Murray and Larry Summers flew from Boston to northern California for final negotiations with Ellison. Summers was in a foul mood. The day before, the majority of the Harvard faculty had voted to censure him for comments made months earlier. At a diversity conference sponsored by the National Bureau of Economic Research, Summers had suggested that "issues of intrinsic aptitude" between men and women might be one reason the latter held fewer tenured positions in science and engineering at top universities. It was a colossally ill-considered thing for anyone to say, much less the president of Harvard, and Summers's position at the university suddenly seemed shaky. Still, the chance to bring a big donation home made him hopeful. "The opportunity was there to mobilize substantial resources," Summers would remember later. Ellison's twenty-three-acre estate in Woodside, California, near Palo Alto, was "insane," Murray thought, modeled on a Japanese emperor's sixteenth-century country residence, complete with a five-acre lake, two waterfalls, and hundreds of mature cherry and maple trees Ellison had planted among the native redwoods.

Getting as comfortable as they could in such formidable surroundings, Murray and Summers agreed with their host that the new venture would be named the Ellison Institute. Its logo would be a tiny globe held in the calipers-like pincers of a large capital letter "E." "Improving world health through accountability" was the tagline. Every major health problem in every country would be studied. Ellison offered his hand to Summers and Murray in turn. They had a deal.

It was a headline grabber—after all, it was the largest gift in Harvard's history—$115 million in initial funding, followed by $50 million annually, beginning in 2009. "The agreement with the university isn't yet signed," *The Wall Street Journal* reported on June 30, 2005, "but Mr. Ellison said in an interview that 'it's absolutely going to happen.'" A comment by Richard Horton followed immediately in *The Lancet*. Under the headline, "The Ellison Institute: monitoring health, challenging WHO," he wrote: "When rumors about the Ellison Institute became known last year, some senior figures at WHO expressed anxiety." Was Murray re-creating the Geneva Evidence and Information for Policy cluster at Harvard? "Murray and Ellison are determined to carve out a niche for an alternative—and better—source of health information than that currently provided by WHO," Horton said. The Ellison Institute would both renew the Global Burden of Disease study and evaluate how specific health programs worked on the ground, structurally independent of governments and aid and advocacy groups.

In mid-June of 2005, Murray and Emmanuela Gakidou married in Athens. The week of their wedding, Gakidou's nephew was thrown underwater while windsurfing. The bride immediately swam out to rescue him. "Chris's mother was pleased," remembers a guest. "She liked strong women." For their honeymoon, Murray and Gakidou went to New Zealand, where they went heli mountain biking. On their return, another couple might have decorated their house and started talking about having children. This pair scouted office space as Murray hired other senior staff members and began advertising around the world for new Ellison Institute fellows. "It was a big thing," says Catherine Michaud. "He"—Murray—"was very enthusiastic."

The plan was to open officially in January 2006 and build up to 130 employees within twelve months. While they waited for Ellison's check, Harvard advanced money to convene expert advisory groups across the United States, Europe, Africa, Asia, Latin America, and the Middle East. Murray was determined to change the world by holding life and death up to a true mirror. With new statistics could come new analysis. With new analysis could come life-saving action. Just as important, he was breaking the lock on basic data. Now public health would be truly public. "Some people at the WHO have no interest in the truth," Murray would say in exasperation. "It's considered sophisticated to be political. There's a pretty quick slope between that and lying." Typical Murray exaggeration, the people he criticized sighed. Typical needless hostility. But who, in the end, would be injured by all that free-floating aggression?

Three months—and then six months—passed. When they weren't off for an Ellison consultation, Murray and Gakidou were traveling repeatedly between Cambridge and Mexico City to complete HIGH's evaluation of the Mexican health system reforms. All the while, their billionaire hadn't sent a penny. "There was agitation and the university wanted to see some money, but we didn't think he was going to renege," recalls Gakidou. Ellison was a friend, Murray thought. They'd gone sailing. He'd welcomed him to his house. Larry Summers was still confident the money would be in hand soon. Using Julio Frenk's office, a king's lair in the beautiful old Mexican Ministry of Health building, which was filled with Diego Riviera murals, Murray spoke regularly with Ellison's lawyers about how the institute would be run and how it would be organized.

One day, the tone changed unexpectedly. Ellison's team said nothing specific, but Murray told Gakidou afterward, "Something's wrong." Within a week, Ellison himself went from commenting on final logo designs to complete silence. It took ten months in limbo to find out why. There was a settlement out of court for insider trading. It required Ellison to make a charitable gift of $100 million. Whether for personal or legal reasons was unclear, but Ellison gave the money to a nonprofit medical foundation he had established earlier. Its emphasis was biomedical research on aging. First, investors said he had cheated them; now, one might observe, he was trying to cheat death.

Could Ellison have given the gift to Harvard if he had so chosen? Murray never knew. In any case, for almost a year, apparently, Ellison debated whether to honor his commitment to Harvard even though he had another nine-figure charge coming. Trying to encourage Murray, one of Ellison's lawyers told him the boss had put the signed payment papers for Harvard in an envelope. All that remained was for them to be dropped off at FedEx. Then the call came to take them out and rip them up. "He'd signed it," Murray says. "That killed me. In some ways, it made it worse."

On February 21, 2006, the other Larry, Larry Summers, announced he would resign as Harvard president. His conflicts with the faculty had never ended. In late June 2006, Summers's last week on the job, a journalist from the London-based *Daily Telegraph* reached Larry Ellison and pressed him to say whether or not he would give what he'd said he would. The answer, definitive at last, was no. "The reason I didn't finish my gift to Harvard was because of the way Larry Summers suddenly left Harvard," Ellison said. "I lost confidence that that money would be well spent." Left unsaid was that he had not responded to any call from Summers since the previous November—or that, two days earlier, the investor Warren Buffett had dramatically upstaged him by pledging $31 billion, roughly double Ellison's entire net worth, to the ever-expanding Gates Foundation. Murray couldn't believe it. "The bottom line is, he welched on his promises," he said.

The Ellison Institute was over. The dream of an impartial, apolitical institute was not going to happen—at least not now, at least not at Harvard. Murray had to let go everyone he'd hired.

All the way across the Atlantic, he could hear the laughter from Geneva.

PART III

Resurrection

CHAPTER ELEVEN

Dinner with Bill

Chumbawamba—"You need some money"—The other side of the continent.

Chris Murray was not someone who shared his feelings easily. After the demise of the Ellison Institute in June 2006, however, it is easy to imagine he felt humiliated. Larry Ellison's decision not to fund the Harvard institute was a personal as well as professional rejection to Murray, and a very public one, covered by everyone from the Associated Press to the *Financial Times*. "At the time it was tragic," Emmanuela Gakidou would say later. "Clinical depression. We'd been traveling for eighteen months." Her voice trailed off. They had found out through the newspaper. "Neither Ellison nor his lawyer had the decency to tell us."

Catherine Michaud, his Harvard colleague, suggests that one of the hardest parts for Murray was having to dismiss the new staff he'd already hired. "Chris was very close with his collaborators," she recalls. When it was over, though, he didn't want to wallow. "He just continued working," Michaud says. You would have had to be with him in the early morning to know how he really felt. Every day at 6 a.m., Murray attacked his elliptical trainer, listening to the Chumbawamba song "Tubthumping": "'I get knocked down. But I get up again,'" he repeated later. "That was my anthem."

It would take more than exercise or anthems to recover from losing Ellison, however. No matter how important the cause of new accountability in public health, it was hard to imagine someone else stepping in with another $115 million check to fund Murray's proposed institute. His idea of somehow competing with the WHO in global health measurement and evaluation seemed to be vanishing in the distance while he was running in place.

The summer of 2006 was an eventful time in other ways at the highest levels of global health. In May, the WHO director-general, J. W. Lee, died suddenly midterm. A record field—thirteen contenders in all, among them Julio Frenk again—were candidates to replace him. In June, nearly simultaneous with Larry Ellison's retreat and Warren Buffett's $31 billion commitment to the Bill & Melinda Gates Foundation, Bill Gates himself announced he would soon step down from overseeing daily operations at Microsoft to focus full-time on philanthropy, with an emphasis on global health.

At first, Gates had envisioned changing the world by supplying free Internet access to public libraries, but an early trip to India convinced him there were more fundamental needs. In 1997, Gates had asked William Foege, a former director of the U.S. Centers for Disease Control and Prevention, how he could learn and then do more about international public health. Foege, skeptical, gave the forty-one-year-old multibillionaire a reading list of eighty-two books. Two months later he and Gates met up again. "I asked, 'How are you doing on those books?'" Foege told *The New Yorker* in 2005. "And he said, 'Well, I have been so damn busy I have read only nineteen of them.' I still didn't know whether to believe him, so I asked, 'Which was your favorite?' He didn't hesitate for a second. 'That 1993 World Bank report was just super,' he told me. 'I read it twice.'"

Gates, of course, meant the 1993 *World Development Report*

containing the first, preliminary Global Burden of Disease findings. It astonished him to discover that a disease he'd never heard of, rotavirus, killed more than 500,000 children annually in the developing world. "I said to myself, 'That can't be true,'" Gates would recall to PBS's Bill Moyers. "You know, after all, the newspaper, whenever there's a plane crashing and a hundred people die, they always report that. How can it be that this disease is killing a half million a year? I've never seen an article about it until now." And this wasn't even an article. "It was just a graph that had, you know, these twelve diseases that kill," said Gates. These included leishmaniasis, schistosomiasis, trachoma—the list of leading scourges, preventable at low cost, whose names he'd also never seen before. "I thought, 'This is bizarre,'" Gates said. "'Why isn't it being covered?' You know, and there's a mother and a father behind every one of these deaths that are dealing with that tragedy."

Gates forced the report on his wife, Melinda, and his father, Bill Sr. Both were equally taken aback. "Something like 'an inactivated polio vaccine' isn't something that rolls off your tongue," Melinda French Gates would say later. "But the idea that some child died because of that disease is something you care about." Especially eye-opening were Dean Jamison's efforts, as editor, to combine Murray and Lopez's new data with his own ongoing cost-effectiveness research. Decades of life could be saved for a few dollars per year per person—but weren't, the Gateses read. They didn't want to be another wealthy couple who gave away millions to things like opera, or art museums, or elite colleges—all worthy, but none with the same life-and-death impact. "The whole thing was stunning to us," Bill Gates told *The New Yorker*. "We couldn't even believe it. You think in philanthropy that your dollars will just be marginal, because the really juicy obvious things will all have been taken. So you look at this stuff and we are, like, *wow!* When somebody is saying to you we can save many lives for hundreds of dollars each, the answer has to be no, no, no. That would already have been done."

All his adult life, Gates had been a believer in Thomas Malthus, the eighteenth- and nineteenth-century English economist and clergyman whose famous *An Essay on the Principle of Population* sternly warned that human populations grow faster than their means of subsistence. Reducing child mortality in the poorest countries was a fool's errand, Gates had thought, because the inevitable result was more people competing for the same limited resources, causing new, worse-yet rounds of war, famine, and disease. The 1993 *World Development Report* told a different story. Overall, the data showed, when childhood mortality dropped dramatically, so did family size. Demographers had first identified the pattern in the late 1920s. The theory was that people want to have enough children so that a certain number survive to an adult age. Therein an apparently paradoxical conclusion: *to cut population, help everyone live longer.* Already, in much of Latin America, North Africa, the Middle East, and East Asia, the number of children per couple was on track to drop from six, seven, or eight to just one or two.

"That is the most amazing fact that should be widely known," Gates would tell Moyers. "You know, essentially, Malthus was wrong. If you raise wealth and you improve health, particularly if you educate women, then this virtuous cycle kicks in and a society not only becomes self-sustaining, but it can move up to a fully developed status."

Multiple times in his business career, Gates had taken the lead in pursuing new ideas: first the personal computer operating system, then the graphical user interface, and, most recently, the Internet browser. Whatever the field, his goal was mastery, and, with it, the greatest possible return for himself and his fellow stakeholders. Now, in 1997, reading and rereading the summary tables of the initial Global Burden of Disease exercise, the CEO saw what Murray saw: a living, breathing world, hidden to everyone without a sharp eye for numbers. "I think there is a general difficulty of looking at a number and having it have the same impact as meeting a person," Gates would say to Moyers. "I mean, if we said right now, 'There's somebody in the next room who's dying, let's all go save their life,' you know, everybody would just get up immediately and go get involved." Much rarer, he observed, was being motivated to do something about "three million kids every year dying of things that are completely preventable with the technology we have today." One sick child was a story. Three million, unfortunately, was a statistic.

In 1999, two years after Gates's life-changing introduction to the Global Burden of Disease, he and Murray met. Murray was in Seattle, trying to raise funds for the WHO, and Gates invited him to dinner at his estate on Lake Washington. At first, the usually hard-to-impress Murray was too starstruck even to process what he was eating. A man whose life's work was built on processing enormous amounts of data was sitting in the home of one of the heroes of the information revolution. Gates, though, was unpretentious and very engaged personally. One extreme numbers enthusiast had met another. The computer software magnate showed the data-driven scientist his personal library of first-edition math books by Leibniz and Newton, and the still rarer and more precious Da Vinci's Codex, which Gates had bought at auction five years earlier for more than $30 million. But arguably the most closely read work in the collection was a dog-eared copy of the 1993 *World Development Report*. Gates practically had it memorized. "He was incredibly wellread and very detail-oriented," Murray would remember later. "He was just stunned—and probably somewhat frustrated—that the empirical basis for what we do in global health was so weak."

Gates was one of the high priests of a culture built around the massive gathering and meticulous analysis of numerical data. Microsoft managers crunched numbers for everything before they made decisions, and they were expected to deliver quantifiable results for each dollar spent. "The idea it takes years to figure anything out—that people didn't know where deaths came from prior to the Global Burden of Disease," says Murray, "he couldn't believe it."

Gates was ruthless, his critics and competitors always said. In fact, he was something rarer: relentlessly rational. In Murray's work, Gates had found the kind of hard data and all-encompassing analysis he demanded at every board meeting. In Gates, Murray had found his ideal reader: a person of imagination and means, committed to doing transformative good, interested in everyone, everywhere, free from preconceptions or political considerations, ready and able to spend freely based on whatever the numbers said would bring the greatest results. The WHO, the best-known authority on global health, had an entire annual budget of only about $1 billion—"the financial resources of a middle-sized university hospital," Gro Harlem Brundtland would write. Gates, by contrast, was worth close to *$100* billion—and he and his wife had decided they wanted to give almost all of it away. The *World Development Report* showed them the right cause, they would come to say. The DALY concept and Global Burden of Disease study showed them the right way. "That began our learning journey," Melinda French Gates remembered fifteen years later. But "not just a learning journey," she continued. "It's 'what can you do?'"

The year before meeting Murray, but directly inspired by the half-decade-old *World Development Report*, the Gateses had given $125 million to found the Bill and Melinda Gates Children's Vaccine Program. Shortly after their dinner with him, they donated

$750 million more to the new Global Alliance for Vaccines and Immunization, or GAVI. Bill Gates later said his hand was shaking with nervousness as he wrote the check. He also said, "This was the best investment I ever made."

According to the 1993 *World Development Report*, "largely preventable or inexpensively curable diseases" of children still caused 43 percent of the burden of disease in sub-Saharan Africa and almost 30 percent of the burden of disease in Asia, excepting China, and in the Middle Eastern crescent. Using such analyses, Gates could be as canny as a philanthropist as he had been as an entrepreneur. "The metric of success is lives saved, kids who aren't crippled," he would say to *Forbes*. "Which is slightly different than units sold, profits achieved. But it's all very measurable, and you can set ambitious goals and see how you do."

In 2000, the Gateses merged all their own previous charitable endeavors under a new name, the Bill & Melinda Gates Foundation, based in Seattle, and endowed—just to begin—with nearly $16 billion. Global health would be by far its largest grant-making area. "Our starting point in deciding where to focus has been the disease burden in developing countries, as measured by disability-adjusted life years (DALYs) lost," the foundation reported. Murray, at this point the WHO's new executive director of Evidence and Information for Policy, was deeply gratified. "That was a big thing," he remembers. "It meant a prime mover in global health was using our metric."

The endorsement was emphatic. Bill and Melinda Gates were giving away their personal fortune, and Murray's equation was their guide. People in the field would speak of the first decade of the twenty-first century as "the golden age of global health," when new funding and attention drove repeated innovations in health care delivery worldwide. The irony was, between 2003 and 2006, at the same time that so many new efforts inspired by burden-of-disease findings were taking off, Murray was pushed from the WHO and his proposed institute at Harvard had to be abandoned.

In June 2006, while Chris Murray was facing the grim task of firing the people he had hired to join him at the Ellison Institute that was not going to be created, his friends Jim Kim and Paul Farmer, once again working together at Harvard Medical School and Partners in Health, were looking for outside support for a new project in Rwanda. *Mountains Beyond Mountains* had highlighted the remarkable results of their medical aid work in Haiti, Peru, and Russia. Now they proposed Rwanda as a laboratory for efforts to improve health systems across sub-Saharan Africa. Murray could help them and they could help him, they believed.

"History, in a very interesting way, brought our projects together almost completely," Kim remembers. He had just returned to Harvard from the WHO after a three-year stint in Geneva as a senior adviser and leader of the organization's HIV/AIDS efforts. In Rwanda, and throughout sub-Saharan Africa, HIV/AIDS was now the second-greatest contributor to the burden of disease, trailing only malaria. Yet the combined burden from noncommunicable diseases—heart disease, stroke, diabetes, mental illness, and so on—was greater still. Murray's data-rich analysis could make the case to invest in building complete health systems that could combat all these problems in a unified way.

A week after news sources around the world reported that Larry Ellison would not, after all, be giving Harvard the money he had pledged, Kim and Farmer had a previously scheduled meeting with Gates Foundation leaders in Seattle. "They were dead keen to bring me along," Murray would recall. He agreed to join them.

The three former fellow medical residents made a pitch to-gether, but each for a separate aspect of health systems. Farmer discussed past successes worldwide and specific opportunities in Rwanda and across sub-Saharan Africa. Kim called for more and better-trained local staff. "Invest in human resources," he said. Murray gave what was by then a two-year-old stump speech about the need for health metrics.

Tadataka "Tachi" Yamada, formerly chairman of research and development at the drug giant GlaxoSmithKline and now leader of international health efforts for Gates, thanked them all for their time. In the end, the Gates Foundation decided not to fund the Rwanda project. Yet Murray's speech resonated. Afterward, as he was sitting down at a restaurant on Second Avenue, Patty Sto-nesifer, the foundation CEO, came up to him. "You need some money," she said bluntly.

It was an observation, not a promise. "There was some hope there," says Murray. But the fact that Larry Ellison was out of the picture was no guarantee that Bill Gates would fund him now.

Despite—or perhaps because of—their obvious commonal-ities as world-changing software entrepreneurs and billionaire businessmen, Ellison and Gates in other ways seemed complete opposites. Ellison would be married and divorced four times by 2010; Gates was a devoted family man who formed his founda-tion's board with his wife and father. Ellison loved extreme sports and distinctive clothes; Gates was a bespectacled bookworm rarely seen in an outfit showier—or more athletic—than an off-the-shelf button-down shirt and khakis. Ellison was tanned and well groomed. Gates's complexion, *The New Yorker* had observed, was characterized by a "definite pastiness," and he "looked as if his most recent haircut had been performed with blunt scissors and a soup bowl." The fellow innovator most closely associated with Ellison was Steve Jobs, the cool-defining Apple CEO and co-founder; Gates, by contrast, was photographed with Warren Buffett, twenty-five years his senior, sitting at a bridge table in Omaha, Nebraska. For Murray, the question was whether he could convince the Gates Foundation leaders that his vision for an independent institute was right even if trying to woo Ellison as that institute's major funder had gone so wrong.

Murray saw Tachi Yamada again in the fall. Almost all major Gates grantees within academia were lab scientists, working on the development of new vaccines. Once more, Murray pushed health measurement. Yamada was very interested, but still unwilling to commit. Or at least not yet.

If Julio Frenk was unsuccessful in his bid to head the WHO, Yamada said, the Gates Foundation wanted to meet with both him and Murray again in Seattle. As it happened, Frenk survived almost three full days of elimination-round voting. When he lost the final November 8 vote, to Margaret Chan, from China, the invitation immediately followed. "That's when we met Bill," Murray remembers.

Murray and Gates had crossed paths at least twice since their initial 1999 meeting, but only in passing. Once was at the Word Economic Forum in Davos. Another was at a social event with Bono. In a sense, though, Murray had already been with Gates for a while. Gates still kept the *World Development Report* as a lodestar. He had funded a partial update of global burden-of-disease numbers by Alan Lopez, Dean Jamison, Murray, and other collaborators, but fundamentally believed global health measurement should be a public sector responsibility. Murray and Frenk agreed: public agencies had their problems, but one lesson of the Ellison Institute debacle was that private money *can't* be held accountable. The problem, they said at the Seattle meeting, was that no one in the public sector was willing to do the job—not the United States, not the European Union, not Japan or China, not the WHO, and not Harvard either unless someone offered another $115 million.

To make a real difference, they argued, global health monitoring would have to take place at an academic research institute. Only then would the work be sufficiently rigorous, and only then would it be considered for the kind of publications that would both validate and disseminate key findings. "Our culture is 'What's the evidence?'" Murray said. "Our goal is to avoid vanity presses and publish our work in the peer-reviewed scientific literature, preferably in top journals." The discipline of having to go through peer review, often with people who weren't happy with what you were saying, would be strengthening. "It doesn't just happen when you submit it," he described. "You're always thinking, 'How will I justify this to a tough anonymous reviewer?'" Such pressure improved every step of the analysis.

"That's why it's so important for us to be part of a university," Murray continued. "The reality is that the WHO has the capacity to publish its own work without peer review and be taken seriously. A lot of what they do never actually gets scrutinized." Most people who thought about global health at all assumed that the WHO was the most reliable source of information on any topic. Certainly they were the best-known. They had the power to publish whatever they wanted, no peer review necessary. And the WHO was good at exactly what Murray was not: getting people to agree. The problem was, as often as not, it seemed that they were getting people to agree on what was convenient to believe rather than what was true.

When Murray talked about the need for more and better health data, he generally found audiences hated too many details. Gates was the opposite. Before the meeting, he had even gone so far as to download and study the latest PowerPoint slides from Murray's Harvard class on global health. "He really goes down into very specific things," Murray noticed. "I've seen him pick out one number from a giant chart and say, 'Explain this.'" The lack of factual details for basic questions like who dies of what still maddened Gates.

What Murray and Frenk were proposing would change everything, they said. The new institute would track major channels of health spending, evaluate key health programs, and, as its flagship product, complete a total Global Burden of Disease revamp, one with the scope of the *World Development Report* and the precision of Mexico's national burden-of-disease studies. Imagine the ability to move analyses from regions like sub-Saharan Africa to individual countries like Angola, and from broad age groups like fifteen-to-forty-four-year-olds to narrow bands like teenagers, Murray told Gates. As people on the ground planned different life-saving and life-prolonging actions, they could see total health loss from any cause, distinguish between death and disability, and watch the results shift over time. Every finding would come with clear links to the specific data sources and estimation methods used, including calculations of uncertainty—how confident the scientists were in each estimate, what they knew well and what required further data gathering. It would be bigger, better, and more comprehensive than anything Murray had led for the World Bank or the WHO and it was all—given funding—within reach.

Gates responded with both encouragement and his own conditions. He liked Murray and Frenk's thinking, and he was satisfied that if anyone could deliver a truly revolutionary new picture of human health, it was Murray. But he had no intention of funding an institute in Cambridge, Massachusetts. To his mind, Seattle was the new capital of global health—at least in the United States. "I'm not giving money to Harvard," Gates said, although he had gone there for two years before ending his college career. If Murray and Frenk wanted to build the kind of institute they were describing, they would have to come to the other side of the continent, to the Pacific Northwest.

Frenk very soon joined the Gates Foundation part-time as a senior adviser. With Murray, the process of finding common ground with Gates took longer. He didn't want an advisory position—he wanted to lead his own independent team of researchers. Again. Since 1992, Murray had headed or tried to head Global Burden of Disease studies at the World Bank, the Harvard Pop Center, the WHO, and the proto–Ellison Institute. This was his fifth and quite possibly final chance to realize his vision. Now, Murray was convinced, nothing less than a completely new institution was necessary if Global Burden was to reach its full potential and be protected from political influence. Maybe, at long last, he had met someone who understood.

Murray, Gakidou, and Michael MacIntyre, HIGH's senior research manager, spent the Thanksgiving and Christmas holidays of 2006 writing their proposals. Thanks to Larry Ellison, they had been polishing the details for close to two years. "All of the accumulated knowledge went into writing the core grants," Gakidou remembers. Would it be enough? Would the foundation sponsor their work?

Murray needed Gates's money to restart Global Burden. Gates needed to be convinced that Murray's analyses were still necessary to invest wisely in health. "Decision makers in health need better information to make effective decisions," Murray wrote Gates. "Information must be comparable, credible, and comprehensible. Methods must be clearly explained and defensible. Debates driven only by emotional advocacy, though compelling, do not always lead to good policy. Inspiration must be matched with information."

By the end of the year, they had submitted their proposals. Weeks later, in late January 2007, the Gates Foundation formally decided to fund a new independent institute, based in Seattle, attached to the University of Washington, led by Chris Murray and called the Institute for Health Metrics and Evaluation. They promised $105 million—contingent on $20 million in additional support from the state of Washington. The University of Washington flew Murray and Gakidou to Seattle for faculty interviews. Soon they both had job offers as professors of global health at the UW School of Medicine and School of Public Health.

By midspring 2007, the state legislature had appropriated the money and the university regents had approved the project. Gates gave Murray's team temporary offices at the foundation's original headquarters, 617 Eastlake Avenue. A new kind of international institution—a local public entity producing a global public good, mainly with private financing—had been formed.

Less than a year had passed between losing Ellison and gaining Gates. But it felt much longer. "There was a window there when the whole Global Burden construct could have died," Murray said later. "He"—Bill Gates—"took time to realize, if he wanted it, he'd have to fund it."

On July 1, 2007, its first day of operation, the Institute for Health Metrics and Evaluation (IHME) had $125 million in initial pledges, approximately $30 million in grants, and three employees.

CHAPTER TWELVE

Risky Business

Going to the Green Zone—A perfect world—Dare to
compare.

When IHME began in 2007, the first task before Chris Murray
was building a new team, now centered in Seattle, capable of
leading the world to better health with a far more detailed Global
Burden of Disease study than ever before. Emmanuela Gakidou
was crucial to the effort, helping to hire staff and creating an
extensive, multiyear fellowship program like those Murray had
started at the WHO and HIGH. Michael MacIntyre would over-
see strategic planning, project implementation, and outside part-
nerships. Together they had to start all over again, beginning with
assembling new groups of workers and new banks of computers,
and finding their way around yet another new home. Their temporary location at the old Gates Foundation headquarters, with
a terrific view of Lake Union, was a far cry from Cambridge.
At first, Gakidou recalled, "I would just stare at the sea planes
landing." The summer passed and IHME moved a mile north,
upstairs from a bakery where she and Murray ate all their meals.

From the beginning, faculty members were interdisciplinary
and international. Haidong Wang, a graduate fellow promoted to
a faculty position as Global Burden's demographer, was Chinese.
The epidemiologist Rafael Lozano, lead author of the original

Mexican burden-of-disease study, headed the effort to determine causes of death worldwide. An American mathematician, Abie Flaxman, designed the software program to model levels of impairment from each illness or injury. Another epidemiologist, Mohsen Naghavi, who had conducted Iran's burden-of-disease studies, would coordinate more than thirty outside expert groups. The biggest, for cardiovascular disease, had one hundred experts in it alone.

Alan Lopez partnered closely with IHME from his position as dean of the School of Population Health at the University of Queensland in Australia, which had its own small burden-of-disease center. The center's director, Theo Vos, a physician born and raised in the Netherlands, had worked across southwest Africa as a rural bush doctor before helping bring burden-of-disease and cost-effectiveness analyses to Mauritius, Australia, Thailand, Vietnam, and Malaysia. Now Vos would lead calculations of years lived with disability for Global Burden. Other essential partners included longtime colleagues of Murray and Lopez at Harvard University, Johns Hopkins University, Imperial College London, the University of Tokyo, and the WHO.

But experienced collaborators were only the beginning. Of the 2 billion deaths since 1970 the new Global Burden would ultimately cover, only about 25 percent had been recorded in a vital registration system accessible to researchers. Understanding the identity and killer of the other 75 percent—some 1.5 billion people—required secondary sources and innovative strategies. "Our core belief is you have to start with looking at *all* the data," Murray said. "And then you can weed it out." His proposal to the Gates Foundation had said the entire project would take three years to complete, giving a deadline of July 2010. Three years to gather and analyze all available details about the health of every person on Earth.

To get the information Global Burden needed, a group of data indexers was established. Their boss, Peter Speyer, a former media executive from Germany, worked the phone like a gossip columnist. "You start cold-calling or network your way to the data set," he explained. Relevant files included countries' hospital and health clinic records, household surveys and census numbers, plus "verbal autopsies," retrospective interviews with family members of the recently deceased. Different countries brought a varied set of challenges. In China, regulations forbade almost all core health data from leaving the country, so Chinese partners had to do analyses and share the results with Seattle. U.S. states, by contrast, sold annual databases of their in-patient hospital users to anyone in the world, for prices ranging from $35 to $2,000. In Ghana, almost the exact equivalent records were available free.

In Nigeria, Africa's largest country by population, the data indexers surveyed hospitals, police stations, health clinics, libraries, colonial archives, and even cemetery plot records. In Libya, the latest census and civil registries turned out to be available online, but only after clicking through seven Web pages written in Arabic. In Iraq, during the end of the American-led occupation, months of spadework revealed the existence of two recent government household surveys. These would help estimate how many Iraqis were being killed or injured by war, as opposed to other causes, a hugely disputed topic. Trying e-mail, Skype, and phone, Speyer finally managed to reach the Iraqi official in charge of statistics and information technology. "She said they'd be happy to share the survey microdata with us, and I said, 'Can you e-mail it or upload it to a website?'" he recalls. "She said no. She burned it onto a CD and told me I had to pick that up in the Baghdad Green Zone."

Speyer really didn't want to buy a Seattle-Baghdad round-trip ticket. Still, this was valuable data. Somehow he would get it. "I had a colleague at the CDC whose sister was working in Bagh-dad," he says. "I asked him if there was a way for her to pick it up." This sister did indeed obtain the CD and mailed it to her brother in Atlanta, who sent it in turn to IHME. Then the translation into English began. "That's two data sets," Speyer observes, "of tens of thousands."

Another completely separate stream of information, and a big one, came from others' published scientific studies. About what? About "health." There were ten thousand articles a month pub-lished with a reference to epidemiology. To the maximum degree possible, Murray wanted all of those results pulled, digitized, and entered into Global Burden, too. Put another way, a fraction of a fraction of the data supplied to the study's scientists was to be everything everyone else had ever discovered.

This was not something that could be stored on desktop com-puters, as in the old days at the WHO. In 2008, IHME moved again to the top floor of a building in Seattle's Belltown neigh-borhood, across the street from the Monorail, midway between the Space Needle and downtown shops and skyscrapers. From his office window, Speyer could see, 4.5 miles away, the University of Washington building that housed the ever-expanding IHME secure supercomputer cluster it was his indexers' job to cram with information. Everything from International Labor Organization injury figures to import and export statistics on asbestos would go into the new study. "The fraction of the population that lives within a few kilometers of a water source is relevant to drowning," Murray explained, as an example. "The number of pigs per capita is relevant for sarcosis." Gathering and organizing each subset of the bigger data was another huge advance in our knowledge of life and death—and just one of dozens of intricate and laborious tasks necessary to complete a single part of the larger study.

Thirty-two research fellows, men and women, all recent college graduates with a talent for data, signed up for two- or three-year stints to help draw the big picture. Their job was to turn all IHME's information into final estimates of death and disability. It was like a Peace Corps for number crunchers. Fellows learned the latest statistical methods, chose a region or health problem, and bore down on the data 24/7 with Murray, Gakidou, Wang, Lozano, Flaxman, Naghavi, and other faculty. One might study AIDS trends in Poland, lung cancer rates in Argentina, diabetes prevalence in Egypt, or the duration of anxiety disorders in South Korea, for example. What killed people? they asked. What made them sick? What was working to save or improve lives? Most earned a masters degree in public health in the process. "The goal is to get the people who would go to Google or Goldman Sachs," Murray said, "but who want to actually make an effect in the world."

Katrina Ortblad, a Dartmouth graduate, was a typically atypical fellow. Once a competitive swimmer, she'd wanted to study art history or design until she realized her visual acuity could be applied as well to health. "It's a stats program, but very few people have a stats background," Ortblad would say of her cohort. Others had been economics, sociology, and anthropology majors. More than half were women. "They all have different styles and angles from which they solve problems," she said. "I feel like I'm half in grad school, half at a consulting company, half at a think tank."

Ortblad's job description added up to a job and a half, which was fitting given the hours she worked. Even when he traveled, Murray still Skyped with her to fine-tune estimates of people with both HIV and tuberculosis. His intermixed concerns and commands, as ever, came rapid-fire: "Something in Western Europe is throwing off all the numbers. Find the regional effect." "Latin America has a huge selection bias in early data. They were only testing people who they suspected had HIV. Just use the last two years." "In East Africa, 70 percent mortality from HIV seems high. So ARV"—antiretroviral—"coverage doesn't do squat?"

Ortblad smiled throughout. "Show him graphs or results," she said, "and automatically"—she snapped her fingers—"he sees the gaps in my data." It became a game in which Ortblad's goal was to find the holes herself first. For every possible approach to a question, she developed hundreds of spreadsheets and graphs for her own reference—her "reserves," she called them (Murray, fondly, termed them "Katrina spam"). "He thinks so quickly and he expects you to think quickly and you want to do a good job," Ortblad explained, justifying her long hours. "If he asks for something, I can pull up the number or graph right away."

If there were statistics IHME didn't have, but needed to know for Global Burden—whether fruit consumption per capita in Bolivia or the fraction of the population riding motorcycles in Indonesia—fellows found the numbers. They conducted literature reviews and coordinated consultations with outside expert groups, traveled abroad for field projects, and dived into the findings on specific diseases, disabilities, or injuries—chronic kidney disease, hearing loss, falls. To estimate levels of intimate partner violence, the team used population surveys and epidemiological studies. How much lunch meat people ate came from nutrition and health research. Thirty categories of injury had to be sorted by both cause (getting hit by a bus, for instance) and nature (head trauma). Well over one *million* sources, published and unpublished, would inform new Global Burden estimates.

One of Ortblad's officemates, Spencer James, applied to medical school, and was accepted, but chose to defer for twelve months to finish his research in Seattle. "The most novel thing about Global Burden is the sense of completeness," James said later. "You have every disease, every country, every age group. To do that, you need these covariates, these predictor variables. We couldn't settle for something that was limited because it would limit all our analyses."

This was perhaps the most impressive aspect of IHME. Global Burden had raised each person who worked there to Murray's obsessive pitch.

Murray gathered an elite team with a set of promises that essentially doubled as demands. Join him and you would contribute to a more important cause than you could find anywhere else. You would work harder than you ever had before. You would push against the boundaries of human knowledge. You would find your limit. A good example was the emerging field of risk factor assessment.

Two major products of every burden analysis were burden by disease and burden by consequence. Knowing the burden by disease told you the scope of a region's health problems—who was sick and dying where, and what they suffered from. Knowing the burden by consequence across the full range of disabilities told you what programs were required to help people get better. For the new and greatly expanded Global Burden study that was going to mark IHME's entry as a world leader in health information and analysis, Murray wanted to add a potentially even more powerful piece of information: global burden by risk factor. This told you root causes: smoking, lack of sanitation, physical inactivity, and so on—the behavior or situation ultimately *behind* each disease, disability, or early death, the wrong moves or unfortunate circumstances that led people to less-than-perfect health in the first place. It was a remarkable assignment—and also a pressing one. Murray had now blown the initial deadline stated to the Gates Foundation. His new goal, communicated to everyone from Gates to staff scientists to the WHO, was 2012.

The way Global Burden would determine the specific risks of any given action or existence was to start from a baseline of no risk at all, an ideal state of being that could exist only in a computer model. Assume drug use (or high salt, or low exercise, or urban air pollution) was *zero*, IHME calculated. Then what would people's health be?

For some things, the baseline was very easy to understand. For smoking, you want no one to have ever smoked. For other health issues, it's less well defined. You can't eat an infinite amount of broccoli. "You can't reduce your blood pressure to zero," Steve Lim, leader of the new risk factor assessment, pointed out, "because you'll die." And what was good for the world's population in general might not be true for each individual: someone who was lactose-intolerant shouldn't be drinking milk, for instance, even if there was evidence that it could lower the risk of certain cancers.

On matters like diet or physiology, the team surveyed and wove together all available scientific literature to identify the ideal average consumption or condition, topic by topic, from trans fats to breast-feeding. On diet, for example, Global Burden took an ideal of eating 300 grams a day of fruit, 400 grams of vegetables, 125 grams of whole grains, and 450 grams of milk. Every week, ideally, an individual should consume at least 114 grams of nut and seed foods, including peanut butter, and no more than 100 grams of red meat. No processed meat (e.g., bacon, salami, and sausages, or deli-style ham, turkey, and pastrami). Certainly no sugar-sweetened beverages, though 100 percent fruit and vegetable juices got a pass. Ideally, infants were to be breast-fed exclusively for six months, then until the age of two as part of a diet sufficient in iron, vitamin A, and zinc. Ideally, only polyunsaturated fatty acids, mainly liquid vegetable oils, rather than saturated fatty acids were to be used in preparing meals. Ideally, everyone ate seafood or supplements sufficient to provide 250 milligrams a day of omega-3 fatty acids. And sodium could not exceed 1,000 milligrams a day, or half a teaspoon of salt.

In this same perfect world, you were also highly active physically. Your household had an unlimited supply of clean water and clean cooking fuels. Radon and lead were absent, the air outside was unpolluted, and nothing at work exposed you to asbestos, arsenic, benzene, beryllium, cadmium, chromium, or a dozen other occupational hazards. You didn't smoke or abuse alcohol or drugs. No child or adult was sexually or physically abused. Your bone mineral density was high, your systolic blood pressure was low, and your body mass index was a perfect 21–23 kg/m.

Everything short of this ideal had specific consequences, cause by cause, which supercomputer models supplied based on IHME estimates of exposure. Breast-feeding, for instance, protected newborns against many deadly infectious diseases and dangerous and painful inflammations of the ear. High bone mineral density helped the elderly recover from falls. Blueberries, carrots, salmon, and safflower oil (to name just four "good" foods) protected everyone against heart disease and stroke. Stopping sexual abuse also led to declines in depression, drug and alcohol use disorders, and intentional self-harm. Cutting diesel engine exhaust forestalled trachea, bronchus, and lung cancers.

All together, the new Global Burden of Disease study would cover sixty-seven risk factors or risk factor combinations for everyone on Earth. "This is all about population level statistics," Lim explained. "It takes into account what is the current consumption relative to ideal consumption." Say everyone in the world ate 300 grams of fruit a day. "Then there would be no attributable burden to fruit. But that wouldn't mean you should stop eating fruit."

One of the increasingly important functions of public health programs is prevention, reaching healthy people who need help knowing how to stay that way. Campaigns against smoking cig-arettes or for wearing seat belts, for instance, save millions of lives without requiring a single prescription. Now, if done right, Global Burden's risk factor assessment could guide new primary prevention programs and public safety legislation. With accurate and complete risk factor information, you could, in theory, *stop any burden before it started* for every people, place, or age.

Of course, you can't really prevent all disease or disability, any more than you can prevent all early deaths. But if you know how much health loss can be blamed on any particular individual ac-tion or condition, and can share that knowledge in a compelling way, you can design interventions that will sharply reduce both personal suffering and medical costs. Was it more important to eat fruits or vegetables, to start exercising or to stop smoking, to rid homes of lead paint or to clear outdoor air pollution? For whom? Where? At what ages?

The new Global Burden would say.

By January 2012, four and a half years after Murray, Gakidou, and MacIntyre had arrived in Seattle, the start of the year in which the new Global Burden study results would absolutely, positively, no excuses, be delivered, the project had grown much larger and become truly global: fifty full-time faculty and staff at IHME; nearly five hundred co-authors in fifty different coun-tries; regular consultations with decision makers on six conti-nents. Even as Murray traveled to Geneva, Washington, D.C., Brasilia, Dhaka, Beijing, Canberra, Auckland, Boston, Atlanta, and Lusaka to share preliminary findings with key public health officials, he and his team raced to process the entirety of their research for final analysis. The Razor scooter had long since been retired as a management tool, and so had the idea of waiting until the end of a project to explain it to everyone. These days, a stranger walking the halls of IHME could observe a color flow chart on almost every desk indicating that person's part of the big picture. Put them together and, like the blueprint for a space telescope, you saw the data and methods necessary to bring the new Global Burden into operation. Still, Murray obsessed over every detail—and kept increasing the study's scope.

In earlier versions, the most sweeping burden-of-disease studies had tracked approximately 100 health problems for one year in eight global regions. In his 2007 grant application to the Gates Foundation, Murray had said the new Global Burden of Disease study would cover 200 diseases and injuries, two different time periods, and twenty-one regions of the world. Now, he wanted to tally 291 ailments and 67 risk factors by age and sex in 187 world countries, charting back over decades. Death's work would be calculated every year from 1990 to 2010, and the swath of illness and injury for every man, woman, and child in 1990, 2005, and 2010. Some estimates, such as life expectancy by age and sex and country, would go back as far as 1970. And the results would be public. It was an open question whether the hundreds of people trying to complete Global Burden could outpace their leader's determined efforts to expand it.

While they were compiling data, Global Burden researchers were also refining their methods. Since Sudhir Anand and Kara Hansen's early Pop Center critique, some four hundred papers had been published on ethical choices in how DALYs were calculated. In July 2011, Murray had convened a meeting of twenty philosophers, ethicists, and economists to discuss the topic. Following their strong consensus recommendation, age weighting (valuing years lived in midlife greater than in childhood or old age) and a related calculation, discounting, were dropped from Global Burden. Among other virtues, the change meant that how

DALYs were determined was even easier to explain to policy makers and the public. Assuming an ideal life span of eighty-six years, "If you die at age ten, you've lost seventy-six years," said Murray. "If you have a disability of 0.2, you've lost 0.2 years."

Skeptics had long attacked the very idea of weighting different disabilities. The international experts whose judgments had formed the initial values did not necessarily represent the general public, they had said. And, anyway, didn't values attached to health vary widely from country to country, and culture to culture? Where people hunt and gather, bad eyesight or a broken leg might be the worst calamity that could befall you. Someone who hunts and pecks on a computer keyboard might place a greater emphasis on avoiding intellectual disability. For those who believe in reincarnation, maybe even death isn't so bad.

Those were powerful arguments, but new surveys that polled much larger and more diverse populations suggested people were much more united in their feelings than anyone had expected. At the Harvard School of Public Health, Josh Salomon, Murray's former employee at the Pop Center and the WHO, completed a wholesale reassessment of the impact of different nonfatal health problems. To determine how judgments varied by region, age, sex, and education level, 220 unique conditions, from asthma to impotence, speech problems to schizophrenia, amputation of an arm to major depression, were compared directly with each other and with dying early, using statistically representative household surveys of the general public around the world. The results were astonishing. "What we found was actually an incredibly high level of consistency across settings," reported Salomon. Correlation between country-specific responses and the pooled model was 97 percent in the United States, for example, 94 percent in Peru and Tanzania, 90 percent in Indonesia, and 75 percent in Bangladesh.

It was difficult to imagine any other topic, from ethics to economics, sex to religion, that would elicit so much global agreement. "There's so much of a common understanding of what health is that transcends cultures," Salomon said later. "I was surprised by just how consistent this is." Africans and Americans alike hated neck pain and were afraid of AIDS; whether Bangladeshi or Peruvian, no one wanted to lose his or her eyesight. Follow-up Internet surveys included people with no education and people with advanced degrees and still saw very little difference.

One might still say only people affected with a condition should be polled on the topic—they alone really know what it's really like. A compelling argument. Yet empirical evidence suggested people with any particular problem almost always rated it *less* severely than people without it. "What's troubling is that we wouldn't want to penalize people because they're good at adapting to a condition and come to the conclusion that that condition isn't worth preventing or addressing," said Salomon. Yes, you can come back from an amputated leg or eating disorder, a stroke or breast cancer, but, as he said, "We don't want to underestimate a condition just because people have a remarkable human capacity to cope with challenges."

Were Salomon's larger conclusion corroborated by other social scientists, it would be epochal. Rich or poor, educated or uneducated, urban, rural, eastern, or western, people generally agree what kind of illness is worse than another, according to his research. The claim that everyone is unique and that they value health states differently is false. Health, put grandly, may be a universal construct. Only in the scope of so enormous an undertaking as Global Burden could this finding be something like a footnote. Now not only how *long* we suffered on average was a matter of data, but how *badly* we suffered, too.

CHAPTER THIRTEEN

Missionaries and Converts

Seventy percent—"Deaths are money"—In the style of Seneca.

lobal Burden was and is a dynamic system. It assumes that every aspect of health affects all the others, and that results will be in a permanent state of flux. From a natural disaster to an infectious disease outbreak to a sudden upsurge in violence, what's worst for a country (or a city, or a family) might be completely different one year to the next. The ultimate goal is always healthy life, but the routes to get there keep changing in ways that have to be constantly calculated and recalculated. As Chris Murray's team grew between 2007 and 2012, a major impetus for Global Burden remained the belief that to know anything, you have to study everything.

To be sure, getting consistent, comparable data on everyone, everywhere was still difficult. But nobody was doing it better than IHME, and many were doing a far worse job of showing the health problems of the entire globe. There were hundreds if not thousands of advocacy organizations in almost every country. Add up all their claims and, as Murray and Alan Lopez had discovered repeatedly since the 1980s, it would be many times the total dead—notwithstanding all the afflicted people around the world whom no one counted at all. If you let advocates sway you without any outside check on their arguments, you would almost certainly sacrifice one deserving group at the expense of another, and hurt as much as heal.

Take the Millennium Development Goals. In the year 2000, all 189 United Nations member states, the WHO, the World Bank, and some two dozen other international organizations had agreed on what they considered the biggest health problems facing the world's poorest people. The selection of problems on which to focus was, in part, political, but the desired outcomes were all expressed in hard numbers. By 2015, they announced, in every country on Earth, we could and should reduce the under-five mortality rate by two thirds and the maternal mortality ratio by three quarters, and we needed to turn back the spread of HIV/AIDS, malaria, and tuberculosis. These and five other resolutions had guided the last decade of these organizations' investments in global health.

"The UN pumps these things out all the time," Murray would say later. "I don't think anyone believed they would have the central role they did." It had been a real accomplishment of the major multinational institutions to focus the whole world on child mortality, maternal mortality, HIV/AIDS, malaria, and tuberculosis. But what percentage of global health loss did those problems really cause? The new Global Burden data showed that 70 percent of the burden globally in 2010 was not related to Millennium Development Goals. And in all of Latin America, all of Southeast Asia, all of East Asia, the overlap was even less.

The Millennium Development Goals were a necessary, vital part of addressing the world's greatest health problems, but they were never sufficient. Now, in part because of all the successful effort in their pursuit, the gap between what the goals did and didn't cover was getting wider. Like similar efforts to reduce infant mortality rates in the 1980s, at a certain point they became their own justification, and became detached from the changing reality of national or global health conditions. By 2010, they didn't even make total sense as priorities in very poor countries where life expectancy was lowest.

Why, for example, treat every nation equally when just *six* countries—India, Nigeria, Pakistan, Afghanistan, Ethiopia, and the Democratic Republic of Congo—now accounted for nearly half of all maternal deaths? And why say that *all* deaths for children under five should decline, but for women target deaths *only* from pregnancy, childbirth, and infectious diseases? In 2000, when the Millennium Development Goals were first established, *maternal disorders caused less than 10 percent of deaths for women aged fifteen to forty-nine* (i.e., their childbearing years). Cardiovascular and circulatory diseases caused 11.1 percent of deaths. Cancers caused 12.9 percent. Suicides, road injuries, and fires caused 11.5 percent.

"Why focus on death of that one cause?"—pregnancy and childbirth—Murray asked. "If you care about the death of mothers, that's ten percent of it. Why not the other ninety?" Mothers everywhere needed a range of health care offerings—cardiologists, oncologists, counselors, and trauma surgeons as well as obstetricians.

Observations like this had found a receptive early audience with program planners at the Gates Foundation, the World Bank, and UNAIDS. But UNICEF's statistics chief wouldn't speak to Murray. And the head of statistics at the WHO, Alan Lopez said, "wishes we would just go away." In estimating total mortality in China, Global Burden results differed from those of the UN Population Division by 20 percent. For central sub-Saharan Africa, the difference was almost 40 percent.

Rivals releasing different estimates didn't bother him, Murray said. What he couldn't stand were those who claimed they and they alone were the authority. One goal of Global Burden was to separate, permanently, science from advocacy. Another was to act as a useful goad to everyone, everywhere, claiming to measure anything in health to get their figures right. It was the men, women, and children served by health programs who mattered to him, and so the statistics that guided those programs had to be gathered and scrutinized as effectively and scrupulously as possible. "Everybody does a better job when they have some notion of being in a competitive environment," said Murray. "If the strength of evidence is there, the arguments will converge. If it isn't, there will be a healthy debate. And that's good for us all."

When Murray started IHME in 2007 and began issuing new reports, he had immediately reignited old controversies in the world of health metrics. With the new Global Burden of Disease study, Murray was again claiming to measure even more than the vast agencies that had been established as part of the United Nations. And Global Burden didn't just challenge UN agencies in the sense of taking over their mandate. It also actively disputed the figures they and other long-standing institutions and organizations had been reporting for years. Even—especially—for the Millennium Development Goals themselves.

For example, for more than two decades, despite a global "Safe Motherhood" movement, the number of women dying annually from complications in pregnancy or childbirth seemed stubbornly stuck at 500,000 or more. Then, in the spring of 2010, armed with new methods and much more data than other studies, the Global Burden team concluded that maternal mortality had in fact dropped by more than a third. As with many other analyses that would form part of the complete Global Burden of Disease study, they submitted these findings for separate publication in advance.

One might have thought fewer mothers dying would be greeted as great news. When *The Lancet* accepted the analysis for publication, however, the editor in chief, Richard Horton, received calls from certain campaigners for women's health urging him to reconsider. "The folks in the community were worried it would change funding or make it seem that they didn't know what they're doing," Murray believed. This is the ongoing predicament that aid workers and advocates find themselves in: their past success sometimes threatens—instead of bolstering—their future ability to continue to reduce suffering or early death.

The flap caused by IHME's analysis made the front page of *The New York Times* on April 14, 2010. As the paper reported, "The findings, published in the medical journal *The Lancet*, challenge the prevailing view of maternal mortality as an intractable problem that has defied every effort to solve it. . . . But some advocates for women's health tried to pressure *The Lancet* into delaying publication of the findings, fearing that good news would detract from the urgency of their cause." Murray, the *Times* continued, described the resistance to his team's report as "disappointing." "It really is an important positive finding for global health," he said. "We believe in the process of peer-reviewed science, and it's the proper way to pursue these sorts of studies."

Independent corroboration came five months later. "Maternal Deaths Worldwide Drop by Third," read the headline of a press release for a fresh report, *Trends in Maternal Mortality*. This time the estimators were those on whose previous figures Murray's team had cast doubt: the World Health Organization, UNICEF, the United Nations Population Fund, and the World Bank. "The number of women dying due to complications during pregnancy has decreased by 34 percent from an estimated 546,000 in 1990 to 358,000 in 2008," the release began. It never mentioned that Murray and his colleagues had said almost exactly the same thing to so much furor earlier in the year, but the new consensus was clear.

In early February 2012, in another Global Burden substudy published in advance in *The Lancet*, IHME concluded that malaria killed twice as many people—1.2 million in 2010—than previously reported by the WHO. To make this claim, researchers had gathered thirty years of data from 105 countries, including new estimates of the effect in Africa of resistance to the most common antimalarial drug, chloroquine, in disease-carrying mosquitos, the availability of a more reliable drug regimen, called artemisinin combination treatment, and environmental factors such as rainfall. Among the new victims they uncovered were hundreds of thousands of adults, which contradicted generations of accepted medical thinking that those who survive exposure to the disease when young acquire immunity for life. People of all ages, they said, needed help.

Tracking the evolution and ecology of a parasitic disease was fundamentally different from counting deaths due to pregnancy and childbirth. But the broad point, stated briefly, was very similar: World Health Organization estimates had been way off. "Malaria Deaths Hugely Underestimated," headlined the BBC. When WHO malaria specialists shot back the same day with their own statement—"key [IHME] findings do not seem to be based on strong evidence"—Murray responded by looking back on what had happened with maternal mortality. "There's a storyline that runs over decades in the global health field, people get used to the storyline, and it's communicated to the public and decision makers," he said. People in the midst of running a specific disease program see any change in their story as a threat. "They react incredibly negatively in the short term. Either they have to spend a lot of effort saying, 'No, no, no, the new story is wrong,' or they have to say, 'Okay, the new story is right, and we have to go back to our communities we've been working in for years, and say *we've* been wrong.'"

The explosive disputes obscured an important fact: the IHME report actually contained lots of good news about beating malaria. Even though Global Burden estimates for the disease's death toll were much higher than the WHO's, the project's researchers said that malaria deaths had peaked in 2004 and declined sharply every year since. And that decline was due to concerted global funding to fight the disease and to new organizations on the ground, including WHO programs such as Roll Back Malaria. The rapid scale-up of insecticide-treated bed nets and artemisinin combination treatment was working, the report noted, and should continue. "We have seen a huge increase both in funding and in policy attention given to malaria over the past decade, and it's having a real impact," Alan Lopez had said at the report's release. "Reliably demonstrating just how big an impact is important to drive further investments in malaria control programs. This makes it even more critical for us to generate accurate estimates for all deaths, not just in young children and not just in sub-Saharan Africa."

The Global Burden leaders believed that the problem was more fundamental than conflicting analyses. In private, they contended that for some of their rivals—"missionaries," Lopez called them—methods didn't matter. These were people who were advocates for a particular group of victims or the fight against a specific disease: nothing was more important than their crusade, and only they could provide the numbers to support it. Ideology trumped evidence. "I don't like missionaries," Lopez said. "They believe that there's one truth and it's theirs."

On May 11, 2012, *The Lancet* published an estimate by an independent child health expert group co-sponsored by UNICEF and the WHO, headquartered at the Johns Hopkins Bloomberg School of Public Health, of 7.6 million under-five child deaths for 2010. According to the Global Burden team, a more accurate count was 6.95 million under-five child deaths in 2010. Whom you believed could reframe a global crisis. Dying children, after all, were commonly considered *the* single worst problem in global health. And, on an individual level, who could argue they weren't? All else equal, every possible resource should be devoted to preventing a death before age five. But IHME believed that the experts tracking such deaths were off by more than 10 percent. *Six hundred and fifty thousand fewer children died in 2010 than experts said.* "There are systematic biases built in," Murray would say later. The Millennium Development Goals, remember, were pegged to progress by 2015. "We are likely to have done better in 2015 than what the world will be pronouncing," he said. "For 2015, we will underestimate achievement of reducing child deaths."

One of the May study's authors, the head of the group at Johns Hopkins, was an official Global Burden partner. There were disputes about who would have final control over the estimates, however, and he had recently stopped sharing data. He had even lodged a complaint with University of Washington leadership, Murray said. There should be only one independent team measuring child mortality, his argument seemed to be. But it was also true that the more children who were said to be dying, the more resources those studying them would get. By following their evidence and lowering overall estimates of child deaths, IHME potentially imperiled research dollars for everyone in the field. "It's just egregious," Murray said. Scientists shouldn't be invested in particular outcomes. "He's an advocate for child health. He knows that deaths translate into money for child health programs. Deaths are money."

This was the kind of inflammatory statement that made it unlikely Murray would ever be embraced unanimously by the community of fellow do-gooders. But the disparities between his and their estimates were too big to gloss over. One hundred and fifty thousand maternal deaths here, 600,000 malaria deaths there, now 650,000 child deaths—the differences were adding up. And the new Global Burden covered not just three ways to die, but *235* causes of death, for each of twenty age groups. Who was closer to the reality on the ground was a professional rivalry with literal life-and-death consequences.

Murray and Lopez would not be muzzled by their critics, and they would not stop challenging accepted truths and widely circulated current estimates. "Who's right?" Murray kept repeating, at IHME and wherever he traveled. "That's the only question. All that matters is being right." Yet for all his intellectual and technological firepower, Murray was once again the upstart, not the established authority. Everyone he contradicted might try to undermine him—and Murray's reliable knack for alienating people might lead him to undermine himself.

In June 2012, the journal *Science* featured an article titled "How Do You Count the Dead?" It was all about the disputes circulating around Global Burden. "Scientists agree they need better estimates for the death toll from the world's major killers," the piece began. "But they fiercely disagree how to go about it."

"There is more at stake than academic reputations," reminded *Science*:

> Global health estimates help determine where billions of dollars in health funding goes. Campaigners use them to justify public health spending on certain causes, such as measles immunization campaigns or AIDS prevention. The numbers also help measure whether a campaign has made any difference, and they are one of the ways policymakers determine whether they are spending their money wisely.

Billions of dollars and millions of lives could shift because of a single study.

Meanwhile, IHME was continuing on as usual—which is to say, at lightning pace. On June 15, Murray and Lopez flew to Seattle together, trading barbs about Lopez checking his bag instead of carrying it on with him. "That's twenty minutes we're not working on Global Burden," Murray complained. Lopez just sighed; he'd heard it all before.

The two had just spent the last four days holed up in Washington, D.C., trying to finalize their findings in time for the annual IHME board meeting. Murray was forty-nine years old now, salt-and-pepper-haired but still slim and boyish-looking. While working, he tapped his toes, aggressive, excited, as if still ten years old and charged with navigating the Sahara. Lopez was sixty, white-bearded but still broad-shouldered, with the bearing of an athlete. Nearly three decades had passed since they had met in Geneva, and two decades since they had begun work on Global Burden. Much had changed in their lives and in the world since then. What was more impressive, though, was what had not. Murray and Lopez still worked closely together toward their shared goal of objectively measuring the health of the entire world. And they still both believed that they could succeed.

Whatever got in the way of improving the data—including food, sleep, and the basic courtesies of workplace life—was something to be jettisoned. They were "Chris" and "Alan" to each other, as they were to everyone at IHME—unless when annoyed or teasing. Then Murray might call Lopez *Dr.* Lopez, and Lopez might call Murray *Dr.* Murray, mocking the other's claim to any sort of exalted status. But Lopez was also about the only person in the world other than Murray's father who sometimes called him "Christopher." This was sweet. "Christopher," he said one day in D.C. when Murray wanted to skip lunch to keep working, "you have to eat."

From the Seattle airport, the pair went to Murray's house,

where they were greeted at close to 10 p.m. by Emmanuela Ga-kidou. As a leading researcher herself, she understood the pressures of big science and took her husband's absences with relative equanimity. "Chris will tell you I'm the least romantic person in the universe," Gakidou said later. "My style is pragmatic, rational. We're well matched." In September 2011, the couple had had a daughter, Natasha. Murray had stayed away from work for a couple of days. Gakidou herself managed maybe three weeks. "You don't live with someone like Chris if you like to go at a slow pace," Gakidou said. Now, approaching midnight, Murray played happily with their nine-month-old. The next morning, he and Lopez were back at work in IHME headquarters.

While Lopez rubbed bleary red eyes, impossibly jet-lagged, and found an empty room to review revisions e-mailed to him overnight, Murray entered his own spacious but sparsely decorated corner office. Half-empty bookshelves held atlases and medical reference texts; scholarly books on politics, health care, philosophy, and economics; a narrative history of the great influenza epidemic of 1918. Atop were small work souvenirs: a clock from the Institute of Medicine, a commemorative disk from the 2011 Indian Census, a decorative bronze urn from the Chinese Ministry of Health, a framed photograph of Murray with former colleagues in Geneva. A colored wood carving on the wall depicted a man on foot leading a desert camel caravan.

In preparation for the upcoming board meeting, a different group from the research team was summoned here every half hour for questions, consultations, exhortations, and abuse. Each scientist showed off his or her own little nervous tic: Steve Lim, in charge of Global Burden risk factor assessments, rubbed his chin; Abie Flaxman, the mathematician turned professor of medicine, bit his nails; Mohsen Naghavi, who led IHME's consultations with outside experts (over the last three years, he'd conducted more than a thousand meetings by phone and Skype alone, he estimated), fingered a string of red beads; Haidong Wang, the demographer responsible for estimating when everyone, in every country of the world, had died, and at what age, for every year since 1970, hugged a Star Wars–branded Moleskine notebook.

Murray, running everyone else ragged, gnawed the end of a whiteboard pen. Special paint had turned the long glass partition that separated his office from the exterior hallway into a six-paned whiteboard on which he and the others marked deadlines, drew graphs, and wrote equations describing analytical problems they were still trying to solve. One pane divided the work ahead into four all-caps columns: "DATA," "ANALYSIS," "REVIEW," "VISUALS."

The point was that every detail mattered. The credibility of the entire study could be undone by a flaw in any of its almost limitless parts. In the days before the presentation to the board, no piece of information was too small to check and check again.

"Something's gone wrong with the numbers," Murray told a risk factor researcher. "There's no way that polycystic ovarian syndrome is forty percent of the female infertility."

"I'll check that," the researcher said.

"If we have cholera in countries that don't have cholera, we'll be killed," Murray said to Rafael Lozano, leader of IHME's cause-of-death analysis group. A 2000 study, he remembered, "had a couple cases of polio in countries that had eradicated it."

"I wrote a letter to the WHO in 2000 because you had a couple cases of yellow fever in Iran," Mohsen Naghavi, a veteran of his country's Ministry of Health and Medical Education, affirmed.

"Exactly," Murray said. "That's what we just can't allow."

With Alan Lopez, Murray peered skeptically at a pie chart. "This says falls and traffic accidents account for 65 percent of all YLDs [years lived with disability] from accidents," he said.

"Suicides are pretty good at it," Lopez observed. "Homicides succeed." Attempted suicides and attempted murder wouldn't show up in disability statistics. "What's left? Fires?"

"I thought animal bites would be higher," Murray said. He e-mailed one of IHME's data analysts for answers. Head tilted, he reviewed updated charts of the relative burdens of neck and back pain, melanoma in Australia and New Zealand, heart disease in Western Europe, and suicide in Africa and Asia. Half an hour later, a new pie chart was in his e-mail in-box, this time with a heretofore missing purple wedge—animal bites—restored to the picture. "That's why we check everything so carefully," said Murray.

Outside his window, tourists rode glass elevators to the roof of the Space Needle. Murray docked his black ThinkPad laptop with an external monitor. Within sixty seconds, he was Skyping with a young IHME fellow, Ian Bolliger, who was sitting in the bedroom of a Seattle apartment indistinguishable from a dorm room. "Where are we with ID?"—intellectual disability—Murray asked him.

"We're going to rerun DisMod"—IHME's disease modeling system—"in all regions," Bolliger said. With his thin beard, black glasses, and poof of light brown hair, Bolliger looked like a DJ, but he had just graduated from Harvard with a degree in applied math. "Abie has a new way to reduce uncertainties."

Murray raised an eyebrow. The last time he had spoken with Abie Flaxman it was to receive news of a bug in his estimates. "Dr. Flaxman should be executed," Murray had said afterward. He was joking, but the humor these days was strained. "Do we know if the rest of your code works?" he asked Bolliger.

"It's half-finished," Bolliger said. He explained how the team was modeling chromosomal and congenital disorders.

"Okay," Murray decided after staring downward, brow furrowed, for several seconds. "That makes sense to me." He looked back at the monitor and noticed an unframed poster on Bolliger's wall. "I like your skiing poster," he said.

"Bode Miller." Bolliger grinned.

Then it was back to business. "When can you get the new numbers?" Murray asked.

Alan Lopez entered Murray's office as the call ended. He carried a printout examined through black-framed reading glasses, low on his nose. "Are there any child deaths in Macau?" he asked.

"There were 150 per 100,000 in 1950," said Murray from memory. Figuring out more recent age-specific mortality rates was the job of a team led by Haidong Wang. And they, too, were not yet finished. "I think Haidong is approaching meltdown," Murray had fretted aloud to Lopez in D.C.

"Keep him alive—we need him," Lopez had said.

Now Wang joined Murray and Lopez in the office. With him was an IHME fellow in her early twenties, Kate Lofgren. Wang, Lofgren, and Lopez waited tensely, shoulders hunched, for Murray's review.

On his computer screen, Murray loaded a PDF document with the team's latest graphs of child mortality estimates over time and, for reference, the income, HIV, and educational attainment rates, of all 187 countries Global Burden studied. Colored circles, triangles, and diamonds marked each data point and its source: census, survey, vital registry, and the like. Red, blue, and black lines—progressively more nuanced IHME models—wove through the points, trying to track the most accurate path. To an outsider it was all very orderly and impressive, but Murray, scrolling quickly, immediately pounced on disparate findings for the years 1970 to 1980 on the document's page 118, Venezuela. "Look at the difference between vital registries and surveys," he said. "Don't add extra variance."

Lofgren, in her second year at IHME after majoring in biology at Smith, nodded. Other young professionals her age talked about software bugs as if they were life-and-death problems. *Word crashed. I lost the whole report. I'm toast.* Her software was actually charting life and death. How many children in Venezuela were dying? Was the situation getting better or worse? How good were official government estimates? What about those by UN agencies? This information, when it was released, would be front-page news.

Murray moved to page 145 in the same document, on Pakistan. His cursor circled the 2008 estimates, a big uptick. "Is this believable?" he said. "If we're saying child mortality is going up in Pakistan, that's a very big deal."

No one answered immediately. "Is that the earthquake?" Murray asked. In 2008, a 6.5-magnitude quake had hit southwestern Pakistan, killing hundreds and leaving 15,000 people homeless, according to immediate news reports.

Lopez cleared his throat. "The trend is clearly going down," he said.

"Get one more data point," Murray told Lofgren and Wang. "We have all these FETP people in Pakistan working with us. We train them. Clearly the numbers exist."

FETP stood for the field epidemiology training program, run for foreign countries by the U.S. Centers for Disease Control and Prevention. As a follow-up to their regular training, twenty-three Pakistanis—doctors, epidemiologists, and government professionals—had since spring been taking an online version of Murray's University of Washington "Global Health Challenges" course. They watched the lectures remotely and answered the same weekly discussion questions as other students. Now, in addition to their four problem sets and final project, Murray had an extra-credit assignment: find new data. "There's the competition between India and Pakistan," he said. "This is very political. We have to get it right."

"I reran the model last night with more runs," said Lofgren. "The picture looks worse."

"How long does it take to run?" Murray asked her.

"Five hours," Lofgren and Wang said in unison.

"I want Alan to see it before he leaves," Murray said.

In front of them, he compared side by side the numbers of their old and new child mortality models, looking at the examples of the Maldives and the Philippines. "The first-stage shift has changed," he noticed.

"I correct bias before anything," said Lofgren.

"That explains what's happening," Murray said. He had found the bug: "Your bias correction has gone wrong."

Lofgren frowned, but in a way that showed she knew how to fix the problem. In all but a couple dozen countries in the world, Global Burden had to project from official but incomplete death records. She'd mistakenly applied the same correction to *unofficial* death records, skewing overall estimates. "It took fourteen hours to run the whole data set," she said, and then huddled with Wang to discuss when to schedule a new run.

"She'll find it," Murray said to Lopez, meaning Lofgren and her now-minor mix-up. Lopez nodded. When Lofgren left, the two men turned together to Wang, who looked miserable to be abandoned. So began a good cop, bad cop interrogation.

"It's a very good paper, Haidong," Lopez said.

Wang exhaled. His shoulders dropped. For the first time in fifteen minutes—perhaps for the first time in weeks—he relaxed.

"Alan has been adding verbiage in the style of Seneca," Murray said. "Lots of dependent clauses."

Wang smiled. A betting person would wager he had no idea what Dr. Murray was talking about. But Wang didn't care. His paper was very good. Dr. Lopez had said so.

"Do you know the differences in the specific country models the UN uses?" Murray asked. On one of his bookshelves, he found and extracted a nine-inch metal dagger from a decorative scabbard. Murray waved it absentmindedly at the demographer.

Wang, blanching, said he wasn't sure.

"We should send you to New York," Murray said.

"They're friendly now, aren't they?" said Lopez, of the United Nations Population Division demographers.

"Friendly-ish." Murray put down his dagger.

Wang thought. "For certain countries," he ventured, "they have a thirty to sixty percent relative difference with UNICEF."

Murray looked to Lopez. This disparity was to IHME's advantage. Given a UN bureaucracy big enough to have dueling numbers, Global Burden could referee. "There's a paper to be written about global mortality estimates by different groups over the last decade," he said.

"I'll raise this with Horton"—Richard Horton, the *Lancet* editor—said Lopez, "because it annoys me immensely that people have been publishing crap."

Murray stared at the sharp point in front of him. They all had to work harder.

PART IV

Going Live

CHAPTER FOURTEEN

Dress Rehearsal

Perfect attendance—Tomorrow's victims—"It's part of the
human condition"—A new agenda—Everything and more.

For two decades, Chris Murray and Alan Lopez had been publishing findings that were part of their larger quest to chart
the entire burden of disease for every place and every person on
Earth. Since it was founded in 2007, the Institute for Health Metrics and Evaluation had been issuing studies that reflected various
parts of the enormous mosaic of global health data being assembled in Seattle. Some, like the revised toll of malaria, had been
extremely controversial. Some, like their advances in data gathering and analysis, had been quietly incorporated into the ongoing
conversation about public health that increased the sum total of
scientific knowledge. But until 2012, nobody outside the institute
had seen the total report—a very inadequate word for a project so
huge and audacious—that was the new Global Burden of Disease
study. Now it was time to put the big picture on display.

Perhaps the most significant prepublication presentation took
place at the annual meeting of the institute's board of directors,
on Thursday, June 21. It was a gathering of global experts at the
highest levels of medicine and international public health. Some
were Murray's former colleagues, some at least occasional rivals. No
one was the least bit inclined to let anything go without a question.

Julio Frenk, no longer with the Gates Foundation and now dean of the Harvard School of Public Health, was the board chair. Other board members included Jane Halton, the secretary of health for Australia; Harvey Fineberg, the president of the Institute of Medicine in Washington, D.C.; Peter Piot, the former executive director of UNAIDS, now director of the London School of Hygiene & Tropical Medicine; K. Srinath Reddy, the president of the Public Health Foundation of India; and Lincoln Chen, Murray's old boss at the Harvard Pop Center, now president of the China Medical Board. Joining them today were leaders of the University of Washington global health department and the UW School of Medicine, the chief of health for UNICEF, the deputy executive director of UNAIDS, the president of the Gates Foundation global health program, and the Gates Foundation's HIV program director and deputy director of measurement, learning, and evaluation. There was also one very interested outside observer in the room. Richard Horton, editor in chief of *The Lancet*, had come to Seattle to see if all the clamor about IHME was the herald of a new order, or just a lot of noise.

After a week of rain, it was the first bright day of summer in Seattle, and the arrivals squinted as they approached the shiny glass-and-steel building that held the institute's offices. One by one, they made their way upstairs and took their places at assigned seats around Murray, his executive team members, and Global Burden lead scientists at the long IHME boardroom table, which sat twenty-five tightly. Almost two dozen more IHME staffers took seats in a second ring of chairs placed against the narrow room's long walls. By the time Rafael Lozano and Haidong Wang entered, a little after 8 a.m., they had to join a handful of IHME fellows dressed up for the occasion and sitting on top of a rear cabinet and side windowsill, jostling a vase of fresh purple orchids, obscuring photographs from Botswana, Tanzania, and

Papua, New Guinea, and interrupting a distant view of Seattle's Elliott Bay.

"Congratulations on achieving one hundred percent attendance," Julio Frenk told everyone at 8:30 a.m. "We have all come with high expectations. You will get a private screening of the Burden of Disease results." Frenk paused. "Chris?"

Murray stood and walked confidently to the front of the room, dressed in what passes for business formal on the West Coast: dress pants, a checked shirt, and a sports coat, no tie. If he was the least bit tired, the institute director did not show it; in fact, he seemed newly invigorated. In the weeks, days, and even hours leading up to this presentation, Murray and everyone else at IHME had been working nonstop to arrive at answers to the most urgent questions about human health and to present them in a way that would be clear, accurate, and persuasive.

To his overseers on the board of directors and at the University of Washington, Murray wanted to establish that he and his colleagues were performing at the highest level as leaders of a gigantic project and heads of a brand-new global institution, and that both the project and the institution would meet or exceed the monumentally ambitious goals with which they'd started together five years earlier.

To representatives of the Gates Foundation, his patron and the most prominent user of previous Global Burden studies, he wanted to prove that he had made good on the foundation's $100-million-plus investment—that the new and improved Global Burden would be completed very soon, that it would offer all the benefits and more he had promised Bill Gates, that the foundation and anyone else working anywhere in the world could use its numbers immediately to save lives, and that they should fund IHME again in the future.

To Richard Horton, his potential editor, he wanted to

substantiate that Global Burden offered not only much more information on the health of the world than any previous scientific study, but also much more accurate information—that it could expand what we knew exponentially *and* correct errors in what had been previously published, including articles that had appeared in *The Lancet* itself.

To his staff, he wanted to show the rewards of all their hard work—how important their colossal project would be, and how close they were to completing it.

To everyone, he wanted to demonstrate that the doubters were wrong—that Global Burden was back, that it was better than ever, and that he and his new team and their five years of unending effort would transform the practice of medicine and public health worldwide in ways impossible to ignore.

The stakes could not have been higher. There were two conclusions that Global Burden made quite clear, each with enormous ramifications for the daily reality of how people lived and died. The first was that humanity's combined efforts in medicine, public health, and global health over the past forty years had been immensely valuable and must continue; real progress was happening, but victory was not yet at hand. The second, more controversial, conclusion was that even at the highest levels of global policy-making, urgent health needs still went unrecognized and effective practices were not being promoted. The big picture in a world with 7 billion people was both inspiring and infuriating.

"Global Burden of Disease," the projected image on the screen behind him read. "The science."

Murray's presentation, co-led by Steve Lim, the head of IHME's new comparative risk assessment work, was a staccato list of jaw-dropping revelations extracted from the masses of data collected over the past five years. For almost four hours, as Murray and Lim talked, the screen behind them flashed with a series of brightly colored charts, a graphic illustration of everything that was wrong in the world. How long we lived, what killed us, and what made us sick had all changed dramatically in recent decades. The details of each of those changes showed where our current course of action was succeeding, and where it needed to improve.

The first big-picture takeaway was that people everywhere were living longer. *Global* life expectancy in 2010—67.5 years on average for men, 73.3 years for women—was about as long as it had been for the *very best-off* in 1970. In Nigeria, the most populous country in Africa, for example, male life expectancy had increased from 47.6 to 58.8; in Brazil, the most populous country in South America, it had gone from 57.8 to 70.5; in China, it had gone from 60.4 to 72.9. The disparities between countries and regions were still vast. But if we all kept gaining at the pace of our predecessors, almost the entire world would soon be living to old age.

The problem with this trend was that health systems and preventive measures had not shifted nearly as fast as the populations they addressed. In countries where the deaths had once concentrated in children, programs now needed to serve young adults. Where young adults had been the primary victims, doctors and health officials could now expect a preponderance of middle-aged patients. And in regions where people in their fifties and sixties had recently crowded hospital and health clinic beds, patients sixty to eighty years old now swelled the rolls.

Not everybody was benefiting at the same rate, however. Look at survival gains not by region, but by age, and it was clear who our rapid health improvements had benefited and who had been left behind. Around the world, Murray reported, children under the age of ten, regardless of sex, were 60 to 70 percent less likely to die in 2010 than they would have been in 1970. In the ages of ten to fourteen, improvement was close to 50 percent. In terms of absolute numbers, the shift from 1970 to 2010 death rates meant almost 20 million child and adolescent lives saved—*a number equal to almost the entire military death toll of World War II, averted.* And not just once. Every year. The WHO, UNICEF, and others should take a bow.

But what happened as they grew up?

In the first four weeks of life—the neonatal period—the overwhelming enemies, unsurprisingly, were those related to birth: being stillborn, or preterm, or asphyxiated, for example. These dangers were largely replaced in the first year by infectious diseases like pertussis, measles, and upper and lower respiratory infections. Between ages one and four, infectious diseases still led the count, followed by nutritional deficiencies and parasitic infections. And in successfully fighting all these causes of death, global health had made great progress. Hence the 60 to 70 percent improvements in survival trends from birth to age five.

Things changed, though, beginning in the five-to-nine- and ten-to-fourteen-year-old age groups. Here, according to Murray's data, the toll from infectious diseases began dropping. What replaced it was surprising. "Intentional injuries"—*violence and suicide*, Murray reported. "Unintentional injuries"—*fires, falls, drowning, poisoning, animal attacks, and other accidents.* "Road traffic"—*walking, riding, or driving.* And in all three areas, he said, "we've made very little progress."

As the rate of injuries rose for older teens and young adults, improvements in life expectancy fell dramatically. This was especially true for men. Between the ages of fifteen and seventy-nine, the average woman worldwide was at least 35 percent more likely to survive in 2010 compared with four decades earlier. For men, particularly men between the ages of twenty-five and thirty-five, gains were as little as 15 percent.

Combined, injuries were responsible for about half of all male deaths and a quarter of all female deaths between the ages of fifteen and twenty-nine. They killed more than 1.2 million men and women in their teens and twenties annually. Each case was tragic. Children whose lives had been saved by vaccines were dying anyway, not much later. And unlike other killers of young adults—pregnancy and childbirth for women, the continuing HIV and tuberculosis epidemics for both sexes—injuries, both accidental and intentional, were a huge threat to health that most policy makers still simply missed.

"This is a good summary of public health globally," Murray said. The higher an age group's improvement in mortality, the more likely that what we were doing now was working. The lower the rate of improvement, the bigger chance that major problems were being neglected or ignored. And injuries were only the beginning of the list.

Between 1990 and 2010, the same chronic conditions that strike people in rich countries, so-called diseases of affluence such as stroke, ischemic heart disease, and diabetes, had become top killers in low- and middle-income countries. In fact, in 2010, according to Global Burden, *two thirds of deaths from noncommunicable diseases occurred in developing countries.* These now caused almost 60 percent of all deaths in these nations, some 23 million lives lost annually. But because they did not kill children and they were not named in the Millennium Development Goals, the new threats had yet to be addressed by international health programs as a whole.

Murray and Lopez had predicted the rising average life span and corresponding changes in causes of death in their early work in the 1980s and '90s. Now that prediction had turned to fact, which meant that health systems still had to take care of all the age groups they had before—and also older age groups with completely different health problems. Central and South Americans needed insulin treatment as well as measles shots. Africans and Asians needed chemotherapy as well as antiretroviral therapy. And people from the Caribbean to the Middle East now required blood-pressure-lowering medication as well as family planning services. Providing these had to be the next frontier of global health. If we didn't respond, Murray made clear, today's survivors would be tomorrow's victims.

It was 9:30 a.m. No one stirred in the tightly packed room, but, already, the implications of Global Burden were stunning. "Why don't we have a UNICEF for men and women, ages twenty-five to thirty-five?" Richard Horton, the *Lancet* editor, would say afterward. "Why don't we have Millennium Development Goals for middle age? I'm not saying we should ignore child deaths, but we should not focus only on child deaths." And in Global Burden, as Murray was about to show, death was only the beginning.

Murray's next topic was the other half of Global Burden: disability, meaning, in this case, all nonfatal health outcomes. Disability was different than death—"what ails you," Murray liked to say, "is not necessarily what kills you." What problems actually caused the most health loss, and to whom?

"As we live extra years," asked Harvey Fineberg, the Institute of Medicine president, "are you going to live them with more or less disability?" What was the impact of rising life expectancy?

Murray answered immediately. Beginning around age five, he said, the average fraction of your life spent sick rose steadily. Between the ages of twenty and thirty, women lost 0.1 years of healthy life to disability annually—about a month a year. Between the ages of forty and sixty, it was closer to two months a year. Then 2.5 months. Then, by age eighty, three months—a full quarter of every year. And the pattern for men was almost identical. The longer you lived, almost wherever you lived, nothing could stop the trend: gain meant pain.

The growing proportion of their lives people spent sick, injured, disabled, or depressed might seem obvious. It was not, at least in health policy circles. James Fries was still being cited for his 1980 claim that the "compression of morbidity" meant ever-shorter periods of poor health for humanity. In fact, the new Global Burden study said, the opposite was true: we were living longer than ever, and suffering ever more diseases and disabilities as we aged. Murray was unequivocal. "What decline we're seeing from communicable diseases, we're seeing a compensatory increase from diabetes." Meanwhile, neurological disorders like Alzheimer's now caused almost twice the years lived with disability than cardiovascular and circulatory diseases combined.

This alone would require major new thinking about health programs and policies, but that was hardly the end of the fresh findings. The unprecedented fine-grained data that Murray and his colleagues had amassed made clear other important results missing or mistaken in previous broad snapshots of global health.

Take sex differences. Women lived longer than men throughout the world, Murray reported, but were in worse health even at the same ages. During 2010, for example, the average forty-year-old man lost 44.5 days of healthy life to disability. A lot. But the average forty-year-old woman lost 48.5 days—four days more in a single year than a man the same age. And the sex gap—data-based, not an opinion poll—was almost lifelong. "Females ten to sixty are one to two percentage points higher than males" in terms of chronic disability, said Murray.

Women had it worse, in other words. They suffered more. And this wasn't in sum, at the end of life. That we could excuse by saying men died early and women survived to be in pain. No, what Global Burden showed was that the average woman hurt more every year for *fifty years*, side by side with the average man of the same age. Hearing this, several members of the IHME board appeared to do a double take.

"There's a way to understand that," Murray said. "Compared to years of life lost"—the things that kill you prematurely—"the key causes of disability"—the things that make you sick—"are completely different." One example was anxiety and depression. Another was neck pain, osteoarthritis, and other musculoskeletal disorders. These in themselves killed no one, but all hurt vast numbers of people very much for long periods of time. Major depression caused more total health loss in 2010 than tuberculosis, according to Global Burden. Neck pain hurt people more than any kind of cancer. Osteoarthritis was worse than natural disasters. And depression, neck pain, and osteoarthritis all struck women more than men the same age.

"Women from ten to sixty have a survival advantage, but a disadvantage in living with disability," Murray concluded. The disadvantage was significant, and not limited to injuries and illnesses related to having children. Women needed more and better care in a variety of areas. With strong campaigns now against the root causes of the global gender gap—biological, social, historical, and economic—they could suffer less and be more productive. They, their families, and anyone they worked with would benefit tremendously—but only if those leading personal, public, and global health programs looked beyond causes of death to imagine healthier lives.

Murray cited specific calculations from the study. In the tiny, wealthy Western European principality of Andorra, for example, female life expectancy was 85.2 years in 2010, among the best in the world. Yet it would be a mistake to believe Andorran women would spend every year of eight decades smiling. In fact, they could expect to lose the equivalent of *sixteen* years of healthy life to nonfatal illness and disability. Subtract time spent ill and their healthy life expectancy was only 69.3 years.

Qatar, Barbados, and Samoa—Murray plucked three more examples at random from the globe. In these countries, female life expectancy in 2010 was 82.1 years, 77 years, and 73.4 years, respectively, a span of about a decade. Yet healthy life expectancy in 2010 for the three groups of women was 65.2 years, 63.3 years, and 62.4 years, a difference of fewer than three years. Put another way, the average Qatari woman could expect to lose one of every five days of her life to being sick or injured.

"We're adding years of life at the point disability goes up exponentially with age," Murray said. "It's part of the human condition. You're going to spend your old age with some form of disease and disability." But if we could identify the worst sources of suffering now, we might be able to combat them.

What were they? Murray listed his team's accounting of the world's top causes of years lived with disability. In the final analysis, number one was low back pain, up an estimated 43 percent between 1990 and 2010 in terms of burden from disability. Second was major depressive disorders, up 37 percent. Iron-deficiency anemia was third, although the burden had actually fallen by one percent. Any adult, man or woman, would see him or herself, friends or family, co-workers or neighbors reflected in one of these conditions or elsewhere in the rest of the top ten: neck pain, up an estimated 41 percent; COPD (chronic obstructive pulmonary disease), up 46 percent; other musculoskeletal disorders, up 45 percent; anxiety disorders, up 37 percent; migraines, up 40 percent; diabetes, up 67 percent; and injuries from falls, up 46 percent. Globally, these were the big misery makers—the largest causes of direct pain and suffering in 2010—and all of them but iron-deficiency anemia were getting worse.

Top 10 Causes of Years of Healthy Life Lost to Disability Wordwide (2010)

CAUSE	ESTIMATED YEARS OF HEALTHY LIFE LOST TO DISABILITY	PERCENTAGE CHANGE BETWEEN 1990 AND 2010
1. Low back pain	57–112 million	43%
2. Major depressive disorder	48–81 million	37%
3. Iron-deficiency anemia	28–62 million	–1%
4. Neck pain	23–46 million	41%
5. Chronic obstructive pulmonary disease	20–42 million	46%
6. Other musculoskeletal disorders	23–32 million	45%
7. Anxiety disorders	19–37 million	37%
8. Migraine	14–31 million	40%
9. Diabetes	14–29 million	67%
10. Falls	14–27 million	46%

These were the ironies of all the gains we'd made, as a species, in pushing back death. First, by definition, we would spend more of our lives with illness because we were going to live longer. Second, the longer we lived, the greater portion of how we hurt was due to middle- and late-age aches, pains, sorrows, handicaps, and bad habits—all largely overlooked by most people in global health. In the near future, years lived with disability worldwide would outnumber years of life lost to early death, Murray predicted. For ever more people, very soon if not already, what made us sick would be worse than what killed us.

"The big surprise to us was back and neck pain," he said. "It has a big effect on people's lives and it's pretty universal." The same was true with depression. And diabetes. And COPD. And falls. If migraines struck you, or you suffered from anxiety, you were not uptight or spoiled; you were a typical human being in pain—and you would be typical anywhere in the world. Even in central sub-Saharan African—Angola, the Central African Republic, Congo, the Democratic Republic of the Congo, Equatorial Guinea, and Gabon—where the leading cause of years lived with disability was iron-deficiency anemia, a close second was major depressive disorder. Third was low back pain.

The leading causes of death often varied dramatically from region to region. Key drivers of disability were much more consistent. "You get a sense of rather stable causes," Murray said. "And these are also the leading causes of money spent in health systems." For all the resources devoted to saving lives around the world, that is, even more of private and public health care costs went to treating nonfatal problems we should instead try to prevent or cure, if possible.

At 10 a.m., ninety minutes after he had begun speaking, Murray finished with the big picture: death plus disability, the number of "disability-adjusted life years" (DALYs) attributable to any health problem. All over the world, from cities to towns to rural villages, a rising percentage of the threats people faced were new to the global health agenda.

Top 10 Causes of Total Years of Healthy Life Lost Wordwide (2010)

CAUSE	ESTIMATED TOTAL YEARS OF HEALTHY LIFE LOST	PERCENTAGE CHANGE BETWEEN 1990 AND 2010
1. Ischemic heart disease	119–138 million	29%
2. Lower respiratory infections	102–127 million	−44%
3. Stroke	90–108 million	19%
4. Diarrheal diseases	78–99 million	−51%
5. HIV/AIDS	75–88 million	350%

6. Low back pain	57–112 million	43%
7. Malaria	63–110 million	20%
8. Preterm birth complications	66–88 million	−27%
9. Chronic obstructive pulmonary disease	66–90 million	−2%
10. Road injury	62–95 million	33%

- -

Ischemic heart disease, up 29 percent in total health loss caused since 1990, ranked first among all causes of global disease burden in 2010. Stroke, up 19 percent, was third. Low back pain, sixth on the list, was so prevalent and so painful it now caused more years of healthy life lost than murder, malnutrition, lung cancer, or tuberculosis. "As the world is aging, the burden shifts," Murray said. Aid programs had to catch up.

None of this was to say old priorities could simply be abandoned. Second of the list of the top ten overall maladies were lower respiratory infections. Fourth was diarrhea. Fifth was HIV/AIDS, seventh was malaria, and eighth was preterm birth complications. Fighting these conditions was still essential. But heart disease, stroke, and back pain were all also major threats that hadn't even been recognized across much of the world. And there were others. COPD was ranked ninth in global burden caused. Road injury was tenth. Major depressive disorder was eleventh. Clearly, diverse advocacy groups had not been wrong to campaign for their respective causes. But now they could be more specific—and, ideally, better coordinated with one another. Realizing the vision of a world with health for all required shifting focus from treating *diseases* to treating *people*, whose particular ailments were always changing.

Murray summed up the findings shortly before 10:15 a.m., clicking to his final slide: a simple black-and-white table showing the fastest-growing causes of the global burden of disease—the new face of public health work everywhere. The many eminent authorities on health in the room stared. Of the health problems increasing fastest worldwide between 1990 and 2010, only one—HIV/AIDS—related to a Millennium Development Goal, but, Murray said, "HIV peaked in 2004." At least six of the top ten—glaucoma, macular degeneration, cataracts, peripheral vascular disease, Alzheimer's disease and other dementias, and benign prostatic hyperplasia—afflicted mainly the aging and the elderly. All were up at least 80 percent over the last twenty years, according to IHME's early estimates. And until now, just about nobody had been paying attention.

I t was 10:15 a.m. Julio Frenk called for a brief recess. While Murray spoke, the atmosphere had been friendly and attentive, though the board and guests had interrupted every few minutes with specific questions. Now the room exploded in chatter, as people parsed aloud the overwhelming consequences of all he'd reported and remaining controversies like the differences between rival child mortality estimates. And, they all knew, there were more controversies to come. When everyone regrouped thirty minutes later, Steve Lim took over from Murray. It was his job to summarize the last and newest major area of Global Burden results—the study's comparative assessment of dozens of health risks. Doctors see many patients they call "the worried well," people who are concerned about ailments they might someday suffer. But public health officials need to do much more about the vast majority of people who don't even know what they could and should be doing to improve their own health. As Lim presented his team's findings, he revealed a repeated mismatch between perceived risks and what was actually worst for us.

Top 10 Risk Factors for Years of Healthy Life Lost Wordwide (2010)

RISK FACTOR	ESTIMATED ATTRIBUTABLE YEARS OF HEALTHY LIFE LOST	PERCENTAGE CHANGE BETWEEN 1990 AND 2010
1. High blood pressure	156–189 million	27%
2. Tobacco smoking	137–173 million	3%
3. Alcohol use	125–147 million	28%
4. Household air pollution	87–138 million	−37%
5. Diet low in fruits	82–124 million	29%
6. High body mass index	77–111 million	82%
7. High blood sugar	78–101 million	58%
8. Childhood underweight	64–92 million	−61%
9. Ambient particulate matter pollution	68–85 million	−7%
10. Physical inactivity	59–80 million	(No data for 1990)

Worldwide, the story was, once more, of the increasing burden of chronic diseases and injuries. In 1990, the leading global risk factor had been being underweight in early childhood, Lim said. The fifth-highest risk factor had been being insufficiently breast-fed. By 2010, as child survival surged, the health loss attributed to both these conditions had dropped approximately 60 percent. Again, advocates for child health could and should be applauded. Now, though, high blood pressure, up 27 percent in total health loss caused, was the leading global risk factor. Tobacco smoking was number two. Alcohol was third, followed by a shocker: *household air pollution*. Hundreds of millions of people in Asia, Africa, Oceania, the Caribbean, and parts of Latin America still used coal, wood, charcoal, and dung for cooking. The practice was more than five times worse for humanity in 2010 than the much more frequently discussed problems of unclean water or lack of sanitation, Lim said.

"This is a good example of how our understanding of causal relationships changes over time," the scientist explained. There was more dirty air inside than we had thought, and this inside air was worse for us than we had known. Like smoking or outdoor air pollution, air from dirty cookstoves turned out to lead to respiratory infections, chronic obstructive pulmonary disease, heart disease, stroke, and cancers. To a smaller but still remarkable degree, household air pollution also put people at risk for cataracts. All ages were affected. At the same time, according to Global Burden, access to clean water, important as it was, was not an all-powerful remedy for what really ailed the world.

Fifth on the list of leading global risk factors in 2010 was another surprise: diet low in fruits. If everyone ate just 300 grams of fruit a day, the study said, it would have improved lives *four times as much as ending all illicit drug use.* The total burden from not eating fruit was so high because a diet high in fruit prevented ischemic heart disease and stroke—the world's two leading killers.

Sixth and seventh were high body mass index, an indicator of obesity, and high blood sugar, a common sign of diabetes, both now worse risks than being underweight as a child. And remember: this was worldwide, not just in wealthy countries, and with an accounting system—years of healthy life lost—intentionally tilted toward the problems of youth. The numbers, and trends, were tied to massive global interventions, or lack thereof. Two decades of interventions against childhood diarrhea and starvation were working, and couldn't stop. But though every case of a sick or starving child was a terrible tragedy, being overweight as an adult, a condition whose attributable health loss had increased 82 percent in twenty years, was now cumulatively worse for humanity. Efforts to remedy obesity had to ramp up.

"Is one of the main messages coming out of this that Africa is a special concern?" someone asked. "The danger is we're excluding Africa here."

"Absolutely," Lim said. *Each* region's risks were different. In eastern, central, and western sub-Saharan Africa, for example, the top three risk factors in 2010 were being underweight as a child, lack of breast-feeding, and household air pollution. Fourth was iron deficiency. Only after that did the subcontinent join the rest of the globe: fifth and sixth were alcohol use and high blood pressure. Southern sub-Saharan Africa, led by South Africa, was a whole different story altogether. *Its* regional risk factor ranking was most like Central America's. First was alcohol use. Second was high blood pressure. Third was high body mass index.

Lim displayed a world map color-coded by the leading risk factor in every country. Like a TV weatherman, he pointed to what Murray termed "the blood pressure belt," contiguous dark green areas through much of Asia to the Middle East where high blood pressure was the leading national risk factor. "This is driven by large salt consumption," Lim observed. He moved to Western Europe and North America, all in orange. Here "tobacco, despite declines, remains the leading risk factor," he said. "In many other countries"—he hopped from Mexico to Morocco, Spain to Saudi Arabia, Fiji to Argentina—"we've seen it be BMI"—body mass index. Alcohol abuse was the leading risk factor in places as diverse as South Korea and South Africa, Belarus and Ecuador. "In Eastern Europe"—including all of Russia—Lim said, "it contributes to a quarter of all mortality."

The risks that Lim were citing were different from everything else the Global Burden of Disease study covered. You didn't just read about smoking, drinking, cooking with charcoal, lack of breast-feeding, or bad diets. You saw them in action: bad choices made everywhere, every day, all around the world. And you knew they involved human behaviors that could change. It was easy to picture a person suffering from malaria, or complications in childbirth, or breast cancer as an unlucky victim, lying in bed, or in a hospital gown, or cradled in the arms of a worried loved one. Risk takers seemed much more like active collaborators with their problems.

But were they? People often speak of the consequences of smoking, drinking, and other dangerous behaviors as "lifestyle" diseases, as though the sufferers actively chose to put themselves at risk. At times that might be true, but often the risk factors that cause disease and disability—anything from limited access to fresh fruits and vegetables to living next to a polluting factory, the stresses of unemployment or a general condition of powerlessness that drive many people to drink or drugs—are beyond the ability of the ordinary individual to change.

In other words, these are risks that cry out for large-scale interventions by governments and aid groups, through programs that are just as challenging and just as important as vaccinating against killer diseases or providing access to clean water. And all of these new strategies and potential gains were going to be part of the new Global Burden reports, Lim said. The study had identified problems that few people in power had recognized as urgent health crises, but once their real risk was seen, they could be targeted for eradication. Charcoal stoves could be replaced by cleaner ways of cooking. Mothers could be better supported in being able to breast-feed. Healthier foods could be made cheaper and more widely available. Unhealthy habits like smoking, drinking, or using too much salt could be discouraged through a number of proven methods. Massive programs of education, intervention, and social action would be needed to change the picture. But it could be done, and it was certainly worth doing. Help change these conditions or behaviors and you could lift the shared burden of death and disability for all.

The presentation finished at noon. For the next half hour, questions for Murray and Lim came quickly from the crowd. Interestingly, the UNAIDS and UNICEF executives both seemed generally supportive of IHME's findings, even when those findings contradicted the reports of their own agency experts. Other board members and guests, including the Gates Foundation representatives, were more probing, particularly of the risk factor work, which Lim acknowledged was in a much earlier stage of development than the parts of Global Burden that Murray and Lopez had been refining for twenty years. Using an iPad, Richard Horton tweeted regularly throughout the discussion, repeating key questions without clearly taking sides. "The GBD is going to challenge in pretty major ways previous estimates *The Lancet* has published," he summarized to followers. "This will trigger important policy debates."

The full significance of the data revealed this morning was still sinking in for everyone. Rio Plus 20, the United Nations Conference on Sustainable Development, was taking place in Rio de Janeiro, Brazil, as Murray and Lim spoke. It was supposed to launch the formation of a new agenda to succeed the Millennium Development Goals from 2015 to 2030. If health officials and organizations accepted the findings of Global Burden, they would need to broaden their efforts far beyond the widely recognized goals of infant and maternal health, clean water, vaccinations, antimalarial bed nets, and the other worthwhile but not sufficient programs already in place. They would have to expand from heart-tugging appeals about little children to more difficult programs targeting heedless teenagers, overburdened adults, and frail elders. "The power of this is so enormous," said Lincoln Chen. "People *care* that their health care systems are not aligned with their national burden. They *care* that it will take a generation to prepare for that. And they *care* that the priorities of governments and donors understand that."

That night, following many more presentations on other aspects of IHME's work, the local and out-of-town eminences gathered at a dinner where the heady praise and buzz about the possibility of reforms to come continued. "What we have is a resource," described Harvey Fineberg, from the Institute of Medicine. "It's both an assessment of disease that's as comprehensive, valid, and reliable as human ingenuity can make it and a platform. Is it a management tool? Yes. Is it a policy tool? Yes. Is it an educational tool? Yes. It's everything and more."

Murray and his staff were not celebrating just yet, however. The dress rehearsal had been an impressive success, but they knew that this was just an early tryout and supporters had outnumbered skeptics. Those at the board meeting were elated to have the new data. But taking IHME's information to the world at large—including its large number of critics—would be the real test.

CHAPTER FIFTEEN

Learning to Swim in the Ocean

A goodwill tour—Translation—"Countries aren't stupid"—
Preparing for the wave—GBDx.

hris Murray and his team were brilliant scientists. But scientists are influential only when their findings are accepted and adopted by others. Now IHME needed to *sell* its new study—to the influential and often critical community of other international health researchers, aid agencies, and charities, and, simultaneously, to local authorities and communities around the world. You could solve humanity's most pressing problems, Murray believed, once you recognized them. That's what got him up in the morning. That's what kept him going all day. "If you don't know what's happening, no one is going to bring their creativity and innovation to address it," he said. "That's why I keep obsessing about marshaling the facts and putting them in a way for people who need to to think about them."

To really make a difference, Global Burden would have to be embraced by both policy makers and the general public. It would have to tell doctors and health officials, city planners and rural midwives, patients, parents, politicians, and all the other people on our big, diverse planet what their worst problems were, how to find the best solutions, and then what progress they (and everyone else) were making in improving their health. Both globally and nation by nation, it would have to help turn information into evidence, evidence into action, and action into results for all.

That was a huge undertaking. The most immediate challenge, though, was just getting people to know the information existed and reassuring them that it was reliable. Even before the new Global Burden of Disease study was published, IHME launched a globe-hopping goodwill (or good data) tour in order to personally persuade directors of public health programs around the world that Global Burden would give them tools they could use to set new agendas and improve their national health systems in transformative ways. To make the most of his extraordinary research, paradoxically, Murray would have to take a break from it. As director of IHME, he would have do what he did worst: make nice. "I'm going to spend two years going to every person in the world, showing them how to use Global Burden," he proposed, only half-joking. "I'm going to start with the Seattle suburbs."

Personally responsible as he felt for Global Burden, Murray was not really the only one spreading the word about it. In early July 2012, for example, Rafael Lozano, leader of IHME's cause-of-death analysis group, traveled to Santo Domingo for an annual meeting of leading health officials from Central America and the Dominican Republic. A Mexico City shoe salesman's son, rotund and philosophical, with a ponytail and bushy mustache, Lozano had a tendency to speak in metaphor. Of a friendly seeming international health official, not to be trusted: "He's like a coin of two faces. False." Of reaching the ultimate decision makers: "You have to ask who is the owner of the circus, and who is working the camels." Of the Global Burden of Disease study itself: "Policy makers like to swim in the swimming pool of data. This tool is trying to teach them to swim in the ocean—and not just at the beach."

The ministers of health from Costa Rica, Guatemala, and the Dominican Republic attended the meeting in Santo Domingo. Belize, Honduras, El Salvador, Nicaragua, and Panama sent high-level representatives. For these "circus owners," Lozano previewed Global Burden's ability to compare burden within and between countries. "I work as a translator," he put it, "from evidence or results to practice."

The comparison mattered, because what harmed people differed widely even within this single region. In Costa Rica, for example, infectious diseases and the problems of childbirth and malnutrition—the usual suspects—were only a small part of the real burden of death and disability. Mental and behavioral disorders, cardiovascular and circulatory diseases, and musculoskeletal disorders each led to greater health loss than all communicable, maternal, and neonatal disorders combined. In nearby Guatemala, by contrast, the leading causes of burden of disease were, one, lower respiratory infections, two, interpersonal violence, and three, diarrheal diseases. "The region is fifty-one million people," said Lozano. "They are eight countries. They share the geography, but they don't share the epidemiological problems."

Within Central America, he identified three distinct subregions. One was countries where health priorities still related closely to the Millennium Development Goals (MDGs), such as Guatemala and Nicaragua. Another was countries where noncommunicable diseases were far more injurious, such as Costa Rica and parts of Panama. Third were countries such as Honduras and El Salvador where both the MDGs and noncommunicable diseases were overshadowed by *violence* as a top cause of health loss. In Central America as a whole, in fact, no single problem stole more years of healthy life from people than homicide—the number of murders here was many multiples of that in the United States, where homicide was in turn a much more serious problem than in Canada, Western Europe, or much of Asia.

"Because Guatemala and Honduras are the biggest countries in the region, and violence is one of their top problems, violence is the top problem for the region," Lozano said. "But this is related to the size of those countries." And only Global Burden also tracked the consequences of *non*fatal conditions—major depressive disorder, ranked fifth regionally, low back pain, ranked seventh, neck pain, ranked eighteenth, and anxiety disorders, ranked nineteenth, for example. "If you look only at mortality, you miss them," said Lozano. "They hadn't thought of that before."

The assembled leaders were excited by the new approach. A burden-of-disease discussion dominated the meeting's question-and-answer session. The ministers and their representatives concluded with a resolution "to understand and put into practice burden of disease analysis at a country level for the region."

Murray, traveling with Alan Lopez and others, arrived ten days later at a summit in Amman, Jordan, where representatives to EMRO—the Eastern Mediterranean Regional Office of the World Health Organization—had gathered. In addition to its headquarters in Switzerland, the WHO had six regional offices: one for Africa, in Brazzaville, Republic of Congo; one for the Americas, in Washington, D.C.; one for Southeast Asia, in New Delhi, India; one for Europe, in Copenhagen, Denmark; one for the eastern Mediterranean, in Cairo, Egypt; and one for the western Pacific, in Manila, Philippines. These had considerable independent organizing power. They might use and promote new

Global Burden findings even if "our cousins in Geneva," as one regional head called his counterparts at WHO headquarters, did not.

"There's this funny thing where in the Middle East, even back to the late 1990s, there's been lots of uptake in terms of the appeals of analytics," Murray noted. "Some parts of the world, you have to sell them on the idea. But not in the Middle East." Representatives came to the meeting with specific technical questions about improving vital registration systems and reducing wrongly coded death records. "There's a bunch of people who went to some of our training workshops, ten or twelve years ago, that have moved up higher in the system," he observed. "They have a bigger understanding. The regional director is on board."

Still, the meeting had two points of tension. First, the Sudanese were upset because Global Burden represented Sudan as one country (the study covered up to 2010; South Sudan split off as an independent state in July 2011). Even anger had its benefits, though, since the Sudanese responded positively by promising to furnish figures that reflected the new situation. "The woman in charge said they would provide us the data we needed as quickly as possible," Murray said.

The other awkward point was sharing the stage with representatives from WHO headquarters in Geneva. These included Ties Boerma, director of the WHO Department of Health Statistics and Informatics, and Colin Mathers, a deputy, originally recruited and trained by Murray and Lopez. "It's hard not to be impressed with it," Boerma would say of Murray's rival team and their study. But the WHO couldn't officially endorse the new Global Burden numbers, even if it wanted to, because, he said, "we have a political dimension." Its member states ran the WHO. Boerma had to consult with them—not to mention other WHO departments and UN agencies—before signing on to any statistics. Global Burden had gone "too fast" for that to happen, he explained. At the same time, rumor had it that if the WHO didn't endorse the IHME-led Global Burden of Disease study, the organization would have no choice but to do its own rival Global Burden. In the future, the WHO might be Murray's client, collaborator, critic, competitor—or all of the above.

Whatever happened, the next few months could represent a tipping point not only in global health but also in global authority. "Countries aren't stupid," Alan Lopez said. "They're going to go where they think they're getting best advice."

More presentations followed: From Amman, Murray flew to Athens; from Athens to D.C.; from D.C. to Seattle; from Seattle to Boston to D.C. again, and then, the plan was, to Brasilia. Until the final trip was canceled, his executive assistant had fretted, "You're in the office five days in July and August."

"Efficiency," Murray answered. He could Skype with colleagues. Friends and family could, if necessary, take a rain check. But there was no way to, say, e-mail the entire Global Burden of Disease study to someone important. Murray had to show up and explain everything in person if he wanted to be heard. Giving national health officials advance notice of the Global Burden results prepared them for the wave of data that was about to roll around the world. And it made them collaborators, local managers of the study instead of surprised bystanders deluged by the flood of new information about to come their way.

In mid-September, he spent forty-eight hours in Riyadh as a guest of the court of the king of Saudi Arabia, co-sponsor of an international conference on healthy lifestyles and noncommunicable diseases. His original itinerary called for a departing flight that left Riyadh at 12:10 a.m. Murray had changed this to a time that was more reasonable, but not by much. "Oh," he said weakly at 7 a.m. the next morning in Cambridge, Massachusetts, his next meeting hub, when asked how he was now. "Dealing with all sorts of Global Burden crises."

He had 3,811 unread e-mails. One contained UNICEF's annual release of new child mortality numbers. "They're at 6.9 million in 2011; we're at 6.9 million in 2010," Murray scanned aloud. The difference, year to year, was now as small as 2 percent. "Each year they get closer to us," he said. He celebrated the growing consensus with a short walk to get a scoop of chocolate ice cream. That night, however, he was felled by stomach problems. He had recently been diagnosed with celiac disease—ironically, one of the few conditions not yet tracked by Global Burden. Given his travels, who knew where he had been exposed to gluten?

Murray staggered home—and, within twenty-four hours, straight back to work. He stayed in Seattle two weeks, enough time for IHME to incorporate thousands of comments from outside reviewers, rerun their analysis one last time, and lock down the numbers they would submit to *The Lancet*. Then Murray went to Mexico City, to a gathering of the people he most wanted, and needed, to have on his side.

The International Association of National Public Health Institutes was the vision of two people: Jeffrey Koplan, former director of the U.S. Centers for Disease Control and Prevention, and Pekka Puska, director-general of Finland's National Institute for Health and Welfare. Koplan and Puska had observed that public health institutes are countries' first responders in detecting, assessing, and addressing major health problems. But many countries with high disease burdens had weak public health capacities, and, in an age of rapidly transmitted infectious disease outbreaks, every country could benefit from increased international collaboration. In 2006, the association had been launched with a first annual meeting in Rio de Janeiro, Brazil, and a $20 million grant from the Gates Foundation. By the fall of 2012, it had seventy-nine members in seventy-four countries on four continents, representing almost 80 percent of the world's people, and their agenda went beyond infectious diseases to a host of new global challenges. On Monday, October 1, 2012, Murray was in rain-washed Mexico City to address them. His stated topic expressed his ambitious goals in the simplest terms: "How to Use Data to Influence Policy."

"This is tricky," Murray said at breakfast. "I've argued about this. My idea of influencing policy, you have to use the media. You have to engage a much broader audience than the power brokers. You don't have a long-term influence on policy unless you infuse public discourse with information. It's no fault of people that they don't know the big picture unless there's some mechanism to provide it."

He had been extremely lucky with Bill Gates, he realized. The world's most successful businessman had recognized that global health was a great *investment*. The outlay of just a few dollars per person directed to the right cause could save lives. But for Global Burden to realize its full potential, and to have the impact that would justify all the long toil, sleepless nights, never-ending travel, and massive financial support, it had to reach everyone else on Earth, too. "A lot of strange ideas people pursue come from a misperception of what priorities are," Murray said. In 2010, for example, rabies killed 50 percent more people than all acts of war, according to IHME numbers. Nearly three times more people died from falls than from brain cancer. At the WHO, he remembered, whenever higher-ups asked for data, a specific news story had compelled their curiosity. Here in Mexico, coverage of medical impoverishment and catastrophic health spending had been crucial to passing Julio Frenk's new national insurance plan. "My experience of ministers of health—unless burden of disease shows up in the media, they quickly forget it," said Murray. "If it shows up, they feel they have to learn about it."

His presentation in Mexico City was not just an introduction of Global Burden's cumulative findings. It was also one of the first live tests of a new way of releasing those findings and making them accessible to the world. In attendance today were the people actually leading programs on the ground to save lives and improve health outcomes in their countries. If you were the head of a national institute of public health, you didn't care about the power struggles between the UN Population Division and UNICEF, the WHO and the World Bank, IHME and other academic centers. What you wanted was new information and analysis to do your job better—to overcome resistance from politicians and deep suspicion from the general public, to increase the well-being and attack the particular health problems of the people you served.

There were two traditional paths to influence public health decision-making, Murray said. "One is scientific: you do a rigorous study, you put it out there, it gets picked up." This was the classic method, and the reason IHME was so focused on getting published in *The Lancet*. "The other is the old way of private conversations in a smoke-filled room. You're relying on a benign dictatorship." But Murray was about to introduce a third method, a twenty-first-century way to share information and reach out directly to a wider audience. To make Global Burden findings completely accessible, IHME's information technology and data development teams had concocted a new online tool, code-named "GBDx," to provide a dynamic platform of constantly updated data that could be consulted by health officials and private citizens alike.

No PowerPoint slides. No spreadsheets. You didn't even have to know what the Global Burden of Disease study was to start using the tool. Just point and click and the software explained itself. The program's original inventor had been an IHME fellow, Kyle Foreman, who started from his own frazzled desire to automate the process of answering Murray's endless queries. This new tool and the country-by-country data behind it would not be available until after the publication of the first round of Global Burden reports focused on larger regions, but Murray wanted to preview it now. If GBDx worked as intended, Global Burden could reach anyone in the world with a Web browser. And IHME could automatically answer questions—What was the leading nondietary risk factor for Estonian adolescents? How many Americans in 1995 died of venomous animal contact? Were Germans or South Koreans more likely to suffer eating disorders?—that even Murray had never thought to consider. Fast, reliable, and useful in an addictive way, it was to be an always-on, always-up-to-date interactive map that tracked where you were in terms of optimal health, where you wanted to be, obstacles along the way, and also where everyone else was at the moment.

Jeffrey Koplan led Murray to a full chamber of the world's public health institute directors. Introducing him was Mauricio Hernández-Ávila, director-general of the National Institute of Public Health of Mexico. "It is a great pleasure to introduce our old friend Chris Murray," Hernández-Ávila told his peers. "He has done a tremendous amount for us here in Mexico."

Before he unveiled the new tool, Murray laid the groundwork. He began his presentation the same way he did everywhere he went, defining Global Burden's methodology and explaining key results. As a rule, the longer-living a region, the more years of healthy life people lost to cardiovascular and circulatory diseases and cancers. Where people, on average, died much earlier in life, burdens were huge for infectious diseases associated with childhood, maternal and neonatal disorders, and HIV and TB. Still, he cautioned, variation between locations could be enormous.

In terms of mean age of death, people in Southeast and Central Asia were almost identical in 2010, for example. Yet, percentagewise, HIV and TB and diabetes were worst in Southeast Asia, while heart problems and cirrhosis hit Central Asia harder. Knowing the difference was important because different regions could benefit most from distinct remedies. Shift from Central Asia to Central America, and the percentage of years of healthy life lost to intentional injuries—violence and suicide—more than doubled. Move from Central America to the Caribbean, and one immediately faced a massive toll for natural disasters, 42 percent of total health loss, the consequences of the 2010 Haitian earthquake.

Condoms, chemotherapy, dialysis, addiction treatment, psychotherapy, disaster relief—every region's most urgent need was different. "Sickle-cell disorders are very variable," Murray said. "Diarrheal diseases, HIV, and malaria are big burdens in developing countries and low burdens elsewhere. Malnutrition is pretty much only in Africa." While lower respiratory infections, ischemic heart disease, and stroke cost vast numbers of years of healthy life in almost every area of the world, poisoning was a high-ranked cause only in Eastern Europe and Oceania. Alzheimer's disease and other age-related dementias, meanwhile, existed as epidemics only in the United States, Canada, Australia, New Zealand, and Western Europe. In Mexico and Morocco, the highest total burden was for diabetes; in Ecuador and Saudi Arabia, it was for road injuries; in Israel and Iceland, nothing subtracted more years of healthy life than low back pain. If time alive and well was our most precious resource, these problems stole the most of it. They were now humanity's worst enemies. For the first time, we knew where they hid, who they struck, how severely, and at what age.

Murray paused. His audience of international leaders looked on intently, immediately interested in how this new information applied to their home countries and the particular causes to which they had devoted their lives. Murray encouraged them all to follow Mexico's lead and produce their own national burden-of-disease studies. Then he urged them to bring the evidence to a larger population than just political leaders. "Fostering a public discussion hasn't been used as much in public health," he said. "You need to communicate directly."

Now was the moment to unveil GBDx. "We've become fully engaged in the idea of dynamic data visualization," Murray said, showcasing the program on the screen behind him. "This is a visual interface. It will go live on our website in 2013."

The starting window showed two panes: above, a series of rectangles for every illness or injury Global Burden studied; below, a world map. Click on any ailment—say lung cancer—and the country colorings on the map below switched to show health loss per capita from that particular cause. Click on any country—say Italy—and the rectangles above resized themselves. The bigger the rectangle, the worse the problem was in that particular place. The darker it was, the more it had increased from 2005 to 2010. And it was all color-coded. Noncommunicable diseases were blue, injuries were green, and communicable diseases and maternal, neonatal, and nutritional disorders were red.*

---

*AA: aortic aneurysm. AFib: atrial fibrillation and flutter. Alzh: Alzheimer's disease. BPH: benign prostatic hyperplasia. CKD: chronic kidney disease. CMP: cardiomy-opathy and myocarditis. Conduct: conduct disorder. COPD: chronic obstructive pulmonary disease. Enceph: encephalitis. FBT: food-borne trematodiases. Glom: glomerulonephritis. HTN Heart: hypertensive heart disease. IBD: inflammatory bowel disease. IHD: ischemic heart disease. Int Lung: interstitial lung diseases. LF: lymphatic filariasis. LRI: lower respiratory infections. MDD: major depressive dis-order. Mech Firearm: mechanical forces (firearm). Mech Force: mechanical forces. Naso: nasopharynx cancer. N Enceph: neonatal encephalopathy. N Sepsis: neonatal sepsis. NMSC: non-melanoma skin cancer. Osteo: osteoarthritis. Oth Circ: other

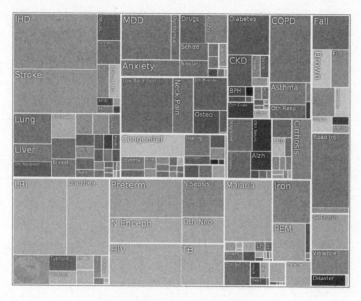

Global DALYs: Both sexes, All ages, 2010

Murray clicked on Mozambique, on the eastern coast of southern Africa. Compared with the global view, red rectangles—communicable diseases—more than doubled in size, from roughly 35 to almost 75 percent of the picture. Move your cursor over the biggest of them, HIV/AIDS, and you learned that the disease had caused 19.5 percent of total burden in Mozambique in 2010. Second was malaria at 17 percent. By contrast, the biggest contributor to

cardiovascular and circulatory diseases. Oth Diges: other digestive diseases. Oth Endo: other endocrine, nutritional, blood, and immune disorders. Oth Inf: other infectious diseases. Oth Musculo: other musculoskeletal disorders. Oth Neo: other neonatal disorders. Oth Neoplasm: other cancers. Oth Neuro: other neurological disorders. Oth NTD: other neglected tropical diseases. Oth Resp: other respiratory diseases. Oth Unintent: other unintentional injuries. Oth Violence: assault by other means. Oth Vision: other vision loss. Parkins: Parkinson's disease. PCO: polycystic ovarian syndrome. PEM: protein-energy malnutrition. PUD: peptic ulcer disease. PVD: peripheral vascular disease. Rheum HD: rheumatic heart disease. Road Inj: road injury. Schisto: schistosomiasis. Schizo: schizophrenia. Sickle: sickle cell disorders. Thalass: thalassemia. V Gun: assault by firearm. V Knife: assault by sharp object. Whooping: whooping cough.

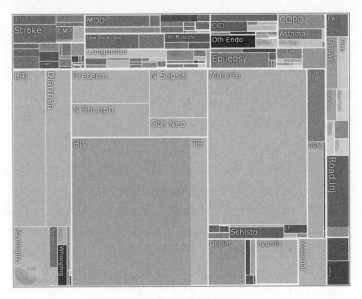

Mozambique DALYs: Both sexes, All ages, 2010

burden worldwide, ischemic heart disease (IHD), had shriveled. In Mozambique, it represented less than *0.5* percent of total health loss.

As Murray demonstrated, one could switch views to see causes of death alone, or just things that made you sick. You could "zoom in" on different types of cancers, injuries from falls versus fires, suicide levels by age group, or death rates over time from, say, meningitis. A separate bar chart broke down country-specific risk factors—iron deficiency, high blood pressure, household air pollution, drug use, smoking—themselves further segmented by the different problems they led to. Diarrheal diseases caused about 1.6 percent of years lived with disability in Mozambique, for example. Hearing and vision loss caused 4.5 percent. A new shading showed what specific risk factors produced each.

The show wowed. Murray had been previewing the new tool here and there, but nobody in the room had ever seen anything like GBDx operating at full power. People nudged each other.

Someone gasped. A man stood and took a picture with his cell phone. Others soon did the same.

Murray kept going. He changed metrics to years of life lost to early death (YLLs) and clicked on the United States. Red causes shrank. Blue causes ballooned. Ischemic heart disease had caused nearly 16 percent of YLLs in the United States in 2010. HIV/ AIDS caused only 1.1 percent. Malaria vanished altogether. From the figures that applied to both sexes, Murray switched to males alone. Injuries—car crashes, self-harm, interpersonal violence, and poisoning—all increased. In 2010, *one third* of years of life lost to early death from interpersonal violence for American men were attributable to alcohol use, the screen said.

"We keep trying to add visualizations as people see this and say, 'I want to see *X*,'" Murray noted.

He toggled to a view of male deaths, ages fifteen to forty-nine, in Colombia. Violence had caused more than 45 percent of the

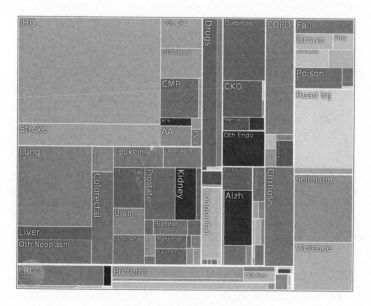

United States YLLs: Males, All ages, 2010

2010 total: 36 percent from firearms, most of the rest from knives. Right behind was HIV/AIDS, causing 15 percent of deaths. The third-biggest killer was road injury. Motorbike riders led that list, followed by car drivers, pedestrians, and bicyclists. Murray moved quickly with the same sex and age group to South Africa. Now HIV/AIDS filled the screen. In 2010, the disease had caused 60 percent of South African adult male deaths before age fifty.

"We've shown this to people with completely no background in health," Murray told the assembled experts. "People ask questions of an informed nature that wasn't really conceivable before when you gave them rows of numbers and tables." He dangled bait. "We believe these are tools you can adopt, tailor to your needs, and make your ability to communicate in countries more available."

Attendees stood to question him. The chief executive of the National Institute for Medical Research in Tanzania said seeing the consistent burden from mental health, country to country, surprised her.

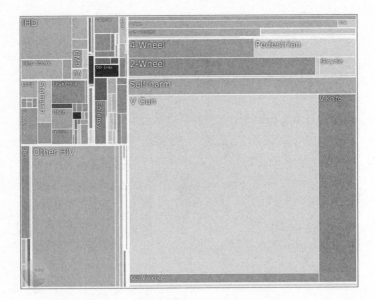

Colombia YLLs: Males, 15-49 years, 2010

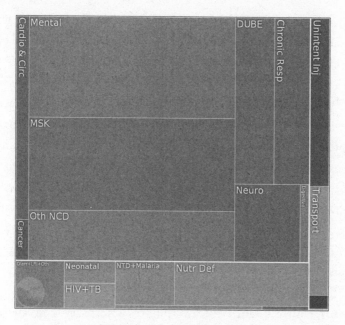

Global YLDs: Both sexes, All ages, 2010

"The burden globally around mental health is equal almost to cancers," said Murray, pointing to two nearly identical blue rectangles. "If I switch the view from burden to disability . . ." He did, and let the picture speak for itself.* The box worldwide for cancers shrank to a wisp; as causes of disability, mental and behavioral disorders swelled to the point they were *forty* times bigger.

"As the data's gotten better, as the analysis has gotten stronger, this fundamental difference between the causes of disability and causes of

---

*Cardio & Circ: Cardiovascular and circulatory diseases. Chronic Resp: Chronic respiratory diseases. Diarr+LRI+Oth: Diarrhea, lower respiratory infections, meningitis, and other common infectious diseases. Digestive: Digestive diseases. DUBE: Diabetes, urogenital, blood, and endocrine diseases. Mental: Mental and behavioral disorders. MSK: Musculoskeletal disorders. Neonatal: Neonatal disorders. Neuro: Neurological disorders. NTD+Malaria: Neglected tropical diseases and malaria. Nutr Def: Nutritional deficiencies. Oth NCD: Other noncommunicable diseases. Transport: Transport injuries. Unintent Inj: Unintentional injuries other than transport injuries.

premature mortality has only gotten clearer," Murray said. "And if we look into the future, the disability part is only getting bigger. We'll have to tackle mental disorders, musculoskeletal disorders, and diabetes."

A man from the Nigerian Institute of Medical Research asked, "Are you going to do a burden for neglected diseases?" Neglected diseases, in the special language of public health circles, meant tropical diseases common only in the so-called third world. These diseases killed approximately the same number of people worldwide in 2010 as ovarian cancer, but, as Bill Gates had noted of his own initial ignorance, most people in rich countries had never even heard of them.

"They're in here," said Murray. "Give me your disease of choice."

"Schistosomiasis," the Nigerian said.

"Of course," said Murray. Out of objectivity and an old-fashioned sense of discretion, he almost never mentioned to audiences his childhood experiences in Africa. But schistosomiasis was the same disease his father had diagnosed forty years earlier in the man Murray found outside their hospital in Diffa, coughing blood. He clicked to select the cause. The map below lit red and orange—high incidence—in Madagascar, Mozambique, Malawi, Nigeria, Benin, Ghana, Liberia, Guinea, and Sierra Leone. According to Global Burden, the total health loss from schistosomiasis in Nigeria was greater than that from hypertensive heart disease in the United States. And the United States had almost 150 million more people.

"Snake bites," the Nigerian parried.

Murray nodded. "We have those under venomous animal bites." He clicked. Now brightly lit were Bhutan, Pakistan, the Democratic Republic of Congo, the Central African Republic, Chad, Niger, Burkina Faso, and Mali.

The man raised his arms. "I give up," he said in a mock surrender that conceded how impressively comprehensive the data were.

Somewhere in the crowd, though, a woman shouted: "Lymphatic filariasis."

"Here it is." Murray picked the tropical disease, the leading cause of elephantiasis, from the same drop-down menu. Worst now were Ivory Coast, Liberia, Sierra Leone, Guinea-Bissau, and Mauritania. "You can ask for thirty- to thirty-four-year-old Afghanis," Murray offered, but there were no takers. He had made his point.

People clapped. Jeffrey Koplan stood. Time was up, but a queue formed to speak directly to Murray. First in line was a pale, stout, balding man in a dark suit, the director of Albania's Institute of Public Health. Second was a woman in a long floral-patterned dress, the head of Sudan's Public Health Institute. Third was a youthful-looking, bespectacled round-faced Ghanaian from the national health service in Accra. "Is there an area of interchange for members who want to work more closely with you and your colleagues?" Koplan asked as a world of health leaders waited their turn to ask questions.

Murray smiled. "That's exactly why I'm here," he said. He looked behind him at the tool. "We want schoolchildren to use it," he told his audience. "We want decision makers to use it. You can ask something no one has ever asked before."

CHAPTER SIXTEEN

London Calling

Early returns—The Royal Society—"He's brilliant, but don't
have anything to do with him."

The Mexico City meeting was a triumphant test of IHME's new
tool to offer all Global Burden findings online in an intuitive,
infinitely customizable format. Before that would be launched,
though, it was essential to be published in a highly regarded scientific journal—one that would validate the Global Burden results
and garner the kind of international media attention that creates
public awareness. *The Lancet* was still their dream publisher, and
Richard Horton, the editor in chief, had been more than enthusiastic at the June meeting in Seattle, but Horton and Murray
had had an up-and-down relationship. *The Lancet* had pilloried
the 2000 *World Health Report*, and every year the journal received about 10,000 submissions. Of these, 175—fewer than 2
percent—were published.

On September 27, 2012, everyone in the IHME office had
gathered when Chris Murray submitted their papers electronically to *The Lancet*. These were global summaries of the country-level assessments: an all-inclusive analysis of death, disability,
and risk factors, for twenty-one world regions. After the boss
hit "send," Steve Lim rang a brass boat bell and some 130 staff
erupted in cheers. Several got teary at seeing the work at last leave

their office. They had all spent weeks, months, and years consumed by the effort. "So many people have been pushing for this to happen," a participant said. "Not just at IHME, but around the world." In total, the papers had 488 named co-authors from 303 institutions in 50 countries.

In mid-October, they heard the judgment from London. The papers were accepted. All of them. Simultaneously. "I am extremely pleased to tell you that we have now received word that seven of the GBD papers have been fully accepted in *The Lancet* and the eighth (the overview paper) will be printed as a commentary with the full paper available online," Murray e-mailed the Seattle team. The result would be an unprecedented *triple*-size issue of the journal. He pointed out that the credit was widely shared: "It has taken a large network of people, both here and elsewhere, to achieve this goal, and it is a testament to your teamwork, diligence, and intellectual drive that we have reached this point. I congratulate you all."

Scientific journals often take months or even years to review and then publish articles. The speed with which the reports of IHME and its collaborators would be published was a tribute to the importance of their research. Eleven weeks after submission, at exactly 5 p.m. on Thursday, December 13, 2012, *The Lancet* would distribute the Global Burden of Disease study. It was the longest issue in the publication's almost two-hundred-year-old history, and the first entirely devoted to just one scientific project. Richard Horton and the entire staff of *The Lancet* had devoted eight weeks to finding reviewers, responding to their comments, and editing the papers. Publishing the new Global Burden was "a landmark event for this journal," Horton wrote in an accompanying comment, "and, we hope, for health."

The day the triple issue came out, Murray, Alan Lopez, and other project leaders were in England to join *Lancet* editors at a press conference at the London School of Hygiene & Tropical Medicine and another briefing for British policy makers at Central Hall Westminster. In person, Horton was even more effusive than he had been in print. "Global Burden provides insights, I think, at the scope and depth of the sequencing of the human genome," he said. "It's the most comprehensive assessment of human health that's ever been published."

Within a few hours, Reuters, *The Guardian*, the Associated Press, Bloomberg News, and Al Jazeera had all interviewed Murray. "People around the world are living longer but with higher levels of sickness," noted BBC News. Forbes.com reported: "Worldwide, hypertension and tobacco smoking are the single largest causes of death and disability." "Deaths from infectious disease are down. Rates of non-transmissible illness—often chronic and frequently the result of obesity . . . are rising," said *The Economist*. Coverage of the findings led *The New York Times* website. "BLOOD PRESSURE: MILLIONS AT RISK," screamed the all-caps front-page headline of the British tabloid *Daily Express*. A photo of Queen Elizabeth II accompanying another story was relegated to the bottom of the page.

Online access to *The Lancet* was free with registration, and a swarm of readers attracted by Twitter posts immediately started combing through the journal, searching the reports for findings on their own specific areas of interest and expertise. Alongside the Global Burden articles on *The Lancet* website were five new IHME-produced interactive visualizations that enabled the viewer to explore global and regional results (GBDx would be released with country-specific data a few months later). Even as Murray stood at an angled podium between two pairs of white columns in Westminster, answering politicians' questions, Peter Speyer, IHME's director of data development, seated a hundred feet away, was working the visualizations and monitoring the traffic on his laptop. A few dozen visitors found the tools in the first minute after 5 p.m. Then one hundred to two hundred. And then *thousands*—to the point Speyer placed a frantic phone call to Seattle. "It's overloading," he warned. "Add capacity."

Rafael Lozano, across the room, watched a scrolling live feed of Twitter users worldwide sharing their own take on the new information. Advocates, aid workers, international officials, and researchers in specific areas leapt to the information about the cause or group they cared most about. "Will mental illness, ranked high in Global Burden of Disease, finally get on global health agenda?" someone asked. "Living longer, but with more chronic diseases," said someone else. "Alcohol single most important cause of death in young adults," reported a third person. "Wow," Lozano whispered, marveling at the diversity and intensity of the interest.

At 6 p.m. Murray left the stage. He crossed the room's plush blue-and-gold carpet, patterned with diamonds and sheaves of wheat. Britain's heroes of science were buried across the street at Westminster Abbey, alongside kings and queens, artists and statesmen honored through the ages. His team's immediate reputation would be more democratically decided by paper citations and website hits. Their legacy would be marked by the lives Global Burden helped improve across the globe.

"How are early returns?" Murray asked Speyer.

"Great," Speyer responded excitedly. "We're scaling up." He checked his e-mail for an update. "Capacity is twenty-five thousand users per second."

Murray fist-pumped. "We're up."

The group regathered that evening with key collaborators and editors from *The Lancet* in a private dining room at the Cavendish hotel. The executive director designate of the Global Fund to Fight AIDS, TB, and Malaria mingled with the scientists, as did the first head of the GAVI Alliance. Bodyguards flanked the special guest Christine Kaseba-Sata, obstetrician, gynecologist, and the first lady of Zambia. "There are very few times in academia you can change the world, and you've done it," Cristian Baeza, the immediate former director for Health, Nutrition and Population at the World Bank, told an IHME researcher.

Murray's wife, Emmanuela Gakidou, arrived. She was accompanied by her parents, in from Athens. Murray hugged them all. Rafael Lozano and his wife were also there. Kelsey Pierce, one of IHME's project officers assigned to help manage Global Burden, brought her stepmother. More than twenty institute fellows and staff would be in town by midnight, she said. IHME had used its accumulated frequent flier miles, so many of them earned by Murray's ceaseless travels, to bring them all to London.

Steve Lim, IHME's chief risk factor researcher, introduced himself to a professor at the University of Cambridge, leader of the analysis of the burden of disease attributable to salt. They had talked to each other on the phone for more than a year, but had never met. Abie Flaxman, IHME's mathematician cum global health professor, chatted with the chair of the cardiovascular disease expert group, a cardiologist born and raised in Ghana. Alan Lopez and his wife, Lene, joined them. Next week they would return to Australia. "I'm there one day," Lopez said. "Then it's three weeks in Argentina, absolutely unreachable."

At 8 p.m. waiters passed out champagne. Murray raised his glass in one hand and the fat new *Lancet* issue in the other. "This is an enormous effort by everyone in this room, and hundreds of others," he told the guests. "The reason so many people here worked on this study—many long beyond what they imagined—is it really does matter." His phone rang, interrupting him. Everyone laughed. "They never stop," said Murray. Then, though, his countenance grew somber. "We have a ritual toast we've made for twenty years," he continued in a voice close to choking. All raised their glasses. "To the reduction of the burden of disease," Murray declared.

The next morning, early rain wet sidewalks outside the head-quarters of the world's oldest continuously operating scientific association, the Royal Society of London, facing The Mall and St. James's Park, a few blocks from Buckingham Palace. Murray entered through a grand door between twin putty-colored columns, past a bust of Charles II, and into a red-carpeted hall lined with heavy drapes, oil portraits of past luminaries, and some 170 black cushioned chairs. All day the *Lancet* publication was being celebrated with an open scientific symposium, five consecutive panels of about six speakers each, covering every part of the study. In three adjacent reception rooms, their inner doors open to make one interconnected space, white-clothed tables held coffee, tea, pastries, and, most important, laptops linked to flat-screen monitors that attendees could use to explore Global Burden visualizations.

At 8:30 a.m. Richard Peto, Alan Lopez's longtime collaborator in estimating global mortality from tobacco, greeted Murray. A decade older than Lopez, Peto—now *Sir* Richard—had wild white hair that made him resemble Andy Warhol, and a famous habit of broadcasting his analyses in progress aloud in a stream-of-consciousness critique. "I've really got to see what I disagree about," he said. Murray beamed. Coming from Peto, anything but direct attack was high compliment. "This is *the* Royal Society," Murray had said giddily three days earlier. "Like Newton. Like Darwin. They presented their results here."

The hall filled. Seemingly every other person here was a Global Burden contributor; among those contributors, outside experts from around the world outnumbered IHME staff four or five to one. They had spent years on the study, too—the list of co-authors on each paper published in *The Lancet* went on for as much as three quarters of a page—but the top-down analysis needed to devise bottom-up global plans of action was all new to them. What was killing us, what was making us sick, and how much were they all changing, they asked, pointing, circling, underlining, and Post-it-noting wherever the region or cause or group they studied was mentioned in printed copies of *The Lancet* they had picked up at the entrance. "We're the people making decisions in where to invest," said a UK National Health Service commissioner. "To see the disability-adjusted life years is fantastic. I see figures on the death end of diseases all the time. But the disability end is where all my costs are."

Discussion continued for nine hours. "Welcome," Richard Horton said from the stage at the beginning. "You've gone through a lot over the last five years. This is a collaboration that is really a complete first." Four hundred and fifty people had attended the symposium in person. Thirteen thousand viewed a simultaneous webcast. "It's been about five hundred minutes from nine a.m. to now," Horton said afterward at 5:45 p.m. "It's dark out again." Based on the 650 million results in the Global Burden of Disease study, he told the crowd, "You've been analyzing 1.3 million outcomes per minute, or 21,000 per second."

While people thronged the reception rooms, congratulating Murray and his team and clustering to plan future publications or points of action, Richard Horton perched in a low-slung chair in a room away from the intellectual fray in order to send a new message on Twitter. "I ended today by learning from several independent sources that WHO directly contacted journalists to trash the GBD," Horton reported. "If this is true, why?"

Science, as though switching reporting teams with the tabloid *Daily Express*, had run a gossipy follow-up to its June article on debates surrounding Global Burden. "Many have questions about how IHME arrived at its results and how they fit with similar efforts by the World Health Organization (WHO), until now the main source of global health data," the story said. The core substantive complaint was that IHME's complex statistical models and computer analyses were a "black box step" other researchers would find very hard to reproduce. But even more of the article was devoted to personal attacks on Murray. "Debates are intensified by some scientists' frustration about what they say is an arrogant attitude . . . ," *Science* summarized. "Murray, widely admired for his intellect and abundant enthusiasm and energy, has come under criticism for his domineering style."

The irony of UN agencies complaining about lack of transparency or institutional arrogance was too much for the editor of *The Lancet*. "The WHO and UNICEF have handled it so badly," Horton said now. "They think they have a God-given right to pronounce on mortality and cause of death and they get very angry if somebody else dares to come along and disagrees with them."

The WHO's director-general, Margaret Chan, had in fact contributed what appeared to be a gracious comment to the *Lancet* issue. In May, she had been confirmed by the World Health Assembly for a second five-year term. The new Global Burden was "an unprecedented effort," Chan wrote. "Accurate assessment of the global, regional, and country health situation and trends is critical for evidence-based decision making for public health. WHO therefore warmly welcomes GBD." Early the next year, she said, "I intend to convene an expert consultation . . . to review all current work on global health estimates."

"They're being totally two-faced and hypocritical," said Horton after the symposium. "They publish a comment—'We welcome it'—but they send out forward troops to criticize and trash the study." And even though multiple WHO staff members had contributed significantly to the Global Burden of Disease study, none had been permitted by their organization to sign their names as *Lancet* co-authors. "They preempt that serious meeting," Horton said, referring to the expert consultation to come, "to say, 'This is a long list about why it's terrible.' No wonder Chris is upset [with the WHO]. I wouldn't have anything to do with them."

Horton was disappointed by the personal sniping, but not really surprised. "Every day I get an e-mail from a scientist that is a very emotional critical response to a paper we've published or a reaction to something they're very angry about," he said. "It's as much about the relationships as it is the science." Researchers were disappointed when *The Lancet* rejected them. But they understood. More than 98 percent of submissions were rejected. What enraged them was when a competitor, or someone whose methods they disagreed with, *was* published, and, on top of that, given such prominence. "It's a very pressured emotional atmosphere," Horton observed. "It's not about rationality at all. It's about priority, it's about getting there first, it's about academic competition. Which is all the bad things about science." He then smiled and shrugged. "But the almost inevitable by-product is all the good things of science." Tomorrow, he noted, he would spend all day at a board meeting for a rival collaboration to the Global Burden of Disease study concerned with tracking maternal, newborn, and child survival. "Chris thinks it's a lot of advocacy and poor science," Horton said. "I totally disagree. I think they have made a massive contribution to child health."

Every great scientist was a missionary, Horton believed. "I don't think there's anything unusual about being activist at all," he said. "Look to the Enlightenment. It's not just knowledge for knowledge's sake. It's for social progress, motivated by this sense of wanting to change things and create a better world, and seeing the power of science to do that." In many fields, science had forgotten its roots. It had become an industry: "for products," said Horton, "for patents." He relished his role in global health because it was observation and experiment, again, as "an instrument for social justice." "That's a very powerful idea," Horton said, repeating, "It's born out of the Enlightenment. We're sitting in the Royal Society. That's what the Royal Society was created to do. Chris does exactly that. That's the Global Burden of Disease."

Horton rose, picking up his red scarf, and checked his phone. It was now 7 p.m. All the wine in the reception rooms had been drained. Alan Lopez and his adult daughter, Inez, talked to Richard Peto. Kelsey Pierce congratulated Peter Speyer on the IHME server not crashing. Katrina Ortblad gave directions to the flat she and nine other IHME fellows were renting in South Kensington for the weekend. They all discussed the institute holiday party scheduled for next Tuesday at the Seattle Aquarium. "It's our last year," said Ortblad. She didn't know what she was doing next.

Murray stood by a portrait of Benjamin Franklin, accepting the day's final well-wishers. "Although he's phenomenally successful, he hasn't had an easy time," Horton said. "He creates a lot of enemies. People misunderstand him." At the time they first met in person, in the aftermath of the 2000 *World Health Report*, "Chris was a hate figure," said Horton. "I was warned off him. 'Yeah, He's brilliant, but don't have anything to do with him. He'll let you down.'" Once he actually met Murray, Horton became much more sympathetic toward him. "He was the opposite of what I'd been told," he remembered. "He was this slightly built, rather vulnerable figure who's incredibly geeky and just wants to talk about data. If he could have worked with the WHO more, he would have."

Horton crossed the room to the subject of his remarks. He asked, "Are you taking a holiday?"

"Two weeks," Murray answered. He paused before adding, "I think." Gakidou, now a few feet away from him, had once told a journalist she "might be able to sit still long enough to spend a couple of days on a beach reading a book." Murray, she continued, "could do it for five hours—at most." Alan Lopez laughed, hearing the story. "I always get a flurry of papers when Chris is in Greece and bored out of his mind," he said.

An IHME fellow said he was going to Switzerland on Saturday. Murray, always eager to be the trailblazer, gave him ski route tips. He gestured dramatically as he demonstrated how to take each cut and pass.

CHAPTER SEVENTEEN

Epic Squared

Landing on the moon—The "aha" moment—"Why hasn't
there been a second report?"

I n the weeks before the Global Burden papers were published in
late 2012, Chris Murray had joked that afterward he was going
to become a man of leisure. Of course, that depends how you define "leisure." Murray and Emmanuela Gakidou's courtship had
been on skis and the couple still "relaxed" with four-hour bike
rides through the Oregon mountains. In 2008, in what might
be considered their homage to Larry Ellison, they had bought
what's called a one-design-class sailboat, the J/105, used in races
where every competitor has the same model boat. "The point is,
it's only your ability to sail," said Murray, who recruited friends
to crew with them and attended seminars on rules and strategy.
"It's a campaign: you have to find your crew, train your crew, op-
timize your resources. Everybody has to do their thing at exactly
the same time." Speed was important. He couldn't tolerate slow
activities, he conceded. Taking care of tiny details that could alter
results were part of the appeal. "You have to get a diver out to
clean the bottom of your boat—a little scum will slow you down."
Following their first race together, Gakidou told him: "That's the
first time I've met people more obsessive than you."

But Murray's obsession wasn't really speed or competition.

What drove him was simple: he wanted to know everything. The more we knew about health, he believed, the more we could improve. "Many people think Chris just likes numbers," said Barry Bloom, the tuberculosis expert, now dean emeritus of the Harvard School of Public Health. "It's absolutely true. He does like numbers. But behind that is a premise that numbers are the only way to hold governments accountable for the health of their people." "What I remember most is the sense of enormity," Alan Lopez said of first tackling the global burden of disease in the early 1990s. "The road didn't exist." He'd alternated between intense feelings of fear and excitement. "It felt like landing on the moon," Lopez said. "What we were doing, I thought, couldn't get done. Then, four or five months in, I thought, 'This is extraordinary. We're going to do it.' I didn't believe it until then. He"—Murray—"did."

The WHO campaign to undermine Global Burden, if it existed at all, didn't work. Neither did the promise to become a man of leisure. After *The Lancet* release of global and regional results, it seemed Murray was everywhere. He visited Geneva and the WHO three times in late 2012 and early 2013 to talk health estimates. UK leaders brought him back to London. He and Bill Gates met together with the prime minister of Norway, one of the largest financial donors to international health efforts among world governments. Top officials at the U.S. National Institutes of Health, the Council on Foreign Relations, the United States Agency for International Development, the World Bank, and the White House all requested presentations. The *Lancet* studies had reported findings by region, but when country-level results came out in the spring, China would have its own Global Burden launch. So would Australia. Brazil was hosting country-specific and regional meetings. And a technical training workshop on burden-of-disease methods was set for Rhodes, Greece, the only country IHME found that was able to approve the international visas of everyone who wanted to come.

Once IHME released the country-level results to the world, Murray intended Global Burden to be a constantly updated, freely available public resource, rather than a once-every-other decade publication. And he wanted to work with partners in individual countries to push the basic unit of analysis from nation to locality. Not just a burden-of-disease study for Europe, Asia, Africa, and North America, in other words, or even for Germany, South Korea, Nigeria, and the United States, but also for Berlin and Bonn, Seoul and Busan, Lagos and Abuja, New York City and Seattle. Think about how much Mexico and Australia, for example, had been able to achieve because of their local burden work. Now the 190-some other countries in the world could catch up. "The global and regional results are really addressed to the scientific community and the donor community," Murray said. "The country results are where we'll really impact policy."

Most of the IHME fellows who had started in 2010 would leave in the summer of 2013, moving on to the next stage of their careers, but new fellows would arrive that fall to work on new iterations of the ongoing story of human health. If the Gates Foundation provided additional backing, as hoped, IHME could also hire a host of new permanent staffers to continue and improve on Global Burden, generating new numbers every year. Murray and his team could then take on measuring everything *else* to do with how we lived and died. "Everything," he put it, "I've been ignoring for so long."

One big new goal was to emphasize the economic significance of health data. Already, IHME tracked not only global health outcomes, but also the spending patterns behind them.

For example, in 2010, noncommunicable diseases in developing countries represented more than 48 percent of their burden of disease. The financial share of health aid devoted to them was *1.2* percent. "Every day, if you show somebody this big-picture view, there's an 'aha' moment," Murray said.

Other measures besides money would also be part of future IHME reports. For instance, the impact of education on health. Past a certain threshold of poverty, the global connection between wealth and health was not really very significant, IHME's analysis showed. What did seem to matter was how many years people spent in school. Vietnam and Yemen had roughly equivalent per capita gross domestic product (GDP), for example, but Vietnamese women averaged 6.3 more years in school, and they were half as likely to die between age fifteen and sixty. Emmanuela Gakidou had studied the education effect—to live longer, get smarter—for more than a decade. "One year of schooling gives you about ten percent lower child mortality rates," she summarized. "Whereas a ten percent increase in GDP—which is huge, China has achieved it for a few years—your child mortality rate would go down only by one to two percent."

The theory behind the observation of this phenomenon was that education produced men and women better informed about their health and welfare and more empowered to seek the best care for themselves and their families. For women in particular, going to school was often the only viable alternative to having children at a young age. The longer women went to school, on average, the older they were before they became pregnant and the fewer pregnancies they had in their lifetime.

To prove the connection between education and health, IHME had had to spend two years gathering academic attainment by age and sex and GDP per capita for every country in the world, 1970–2010. Getting the information had been a dogged struggle—another "impossible" project IHME was determined to show could be done. Even for GDP, "across all countries, all years, about half the data points were missing," Gakidou said. "What was there was not always correct." Soon they'd add measures of income inequality *within* nations.

"First you did the work the WHO was supposed to do," Peter Piot, the IHME board member and director of the London School of Hygiene & Tropical Medicine told her and Murray. "Now you're doing the work the World Bank is supposed to do."

Another ongoing project involved checking and comparing the claims of major international aid groups. IHME's sponsors and Global Burden's earliest supporters came under scrutiny along with everyone else. They, too, committed the sin of self-assessment, with all its attendant inaccuracies. The Gates Foundation funded GAVI, formerly the Global Alliance for Vaccines and Immunization, and the Global Fund to Fight AIDS, TB, and Malaria, for example, and "[They] and the World Bank have been making 'lives saved' claims," Murray said. Every aid group did—these were just the biggest. He pulled up clippings from the organizations' websites: "6.5 million lives saved through Global Fund–supported programs"; "GAVI support has prevented 5.4 million future deaths"; "13 million lives saved through IDA [World Bank international development assistance] during the last decade." The same kind of double- and triple-counting he and Lopez had found before Global Burden was still common. "GAVI buys vaccines," Murray outlined. "They claim the full health benefit of every kid who gets a vaccine. The World Bank says, 'No, no, we trained the health care workers and we pay for the refrigeration.' You need to track investment in outcome across different value chains."

International institutions like GAVI, the Global Fund, and the World Bank, important as they were, were only part of the puzzle. Even with billions of dollars in donor backing, national governments in the developing world were spending an average of twenty times the amount of outside aid on their own health programs. What they didn't have, though, was good information on how the money flowed, or whether what they were doing was working. In other words, they lacked full understanding of their health systems. How much did it cost to provide a country's health services? Who paid for and who received them? What were the largest barriers to access? Which populations were the most affected? Those had been the kinds of questions asked and answered by the 2000 *World Health Report*, the next topic on IHME's to-do list. The project was called ABCE, with each letter standing for one key study area: Access, Bottlenecks, Costs, and Equity.

"The project overall is fairly ambitious," Emmanuela Gakidou noted in a typical understatement. "In Global Burden, you borrow strength [of good data] across regions. There are similar epidemiological profiles. In health systems, you can't borrow strength." Canada's health system is different from the United States'. Colombia's is different from Mexico's. Zambia's is different from Uganda's. "*Each* country is totally different," Gakidou explained. "You can't extrapolate."

Differences were as large or larger within countries as between them. She listed some of these differences: "great facilities, no facilities; great equipment, lousy equipment; huge variations in wealth." In a mountainous nation, barriers to care might be geographic. In another location, the constraints could be cultural. For example, Gakidou observed, if no one spoke their language, "indigenous women might not go to a public clinic." It wasn't that reducing inequalities should always be the top priority, "but," she continued, "you can't consider them unless you measure them. You should always have them in mind when you make decisions.

The idea is to find ways to reduce the inequalities that are amenable to change."

The growing reach of IHME's data was stretching in still other directions, gathering information on how well health programs were really operating and how doctors and patients on the local level could improve outcomes. For individual countries and aid programs, IHME would evaluate the real-world coverage of different interventions: contraceptives, insecticide-treated bed nets, childhood immunizations, new maternity wards, and more. Steve Lim had already started traveling back and forth between Seattle and Zambia, Uganda, Yemen, Mozambique, and India, on a four-year, $16 million evaluation project for GAVI. The level of detail was such that his team was double-checking hospital supply cabinets to see what was stocked, sticking their own thermometers in vaccine storage sites to make sure dosage effectiveness wasn't compromised by bad equipment, and interviewing patients afterward about how they'd been treated. Thousands of facilities were covered. "It's fundamentally asking, what works and what doesn't?" Lim said. "That's what's exciting. It's both process and impact."

In May 2012, even as everyone was still obsessed with completing Global Burden in time for the board of directors meeting in June, IHME had signed a $10 million agreement to bring a supersized version of its health tracking system to the entire nation of Saudi Arabia. The lead scientist on the project would be Ali Mokdad, who had grown up and gone to college in Beirut, Lebanon, but had joined IHME as a faculty member after two decades at the U.S. Centers for Disease Control and Prevention, including the position of chief of the behavioral surveillance branch at the National Center for Chronic Disease Prevention and Health Promotion. "Every health information system in the world, including in the United States, is developed in silos," Mokdad said. "We should know in a country what exactly people go to hospitals for. We should know what is done for everyone who has a heart attack. How as a society are we reacting to an increase in cardiovascular disease?" The Saudis offered virtually unlimited funding to try to put all the data in their country together.

Devoting resources to information gathering was essential to the success of any public health system, Mokdad knew, but it had a public relations problem: if it worked well, *nothing happened*. "If I'm a minister of health and I decide to do it, it's expensive and it takes four or five years to be fully functioning," he explained. "I can't show I'm doing something right now. Even worse, I can't take credit. People can't see surveillance, but everyone can see me doing a ribbon cutting at a hospital." Democratically elected representatives, hereditary rulers, and military strongmen share this common bond: they all know that appearances matter.

The case he made to politicians was that no leader could afford an outbreak where hundreds of people died of a preventable disease. "Say in a country, part of the water coming to a school was contaminated with sewage," Mokdad said. "Three kids drink it and die of diarrhea." The surveillance system would flash red: outbreak. "You investigate and then you find out what happened," said Mokdad. "You shut the water source down, you add chlorine, you fix it. Your glory is, yes, three children died, but it could have been many more."

Not all the new information IHME planned to make available was far away or in a developing country. Some was very close to home. "The U.S. is an example of getting really lousy outcomes for our income level," Murray said. "That's an undisputed fact. The reasons are very disputed." How much of poor health for some Americans was due to bad habits? How much due to income level? How much because of lack of access to medical care or inadequate treatment? No one really knew. Ask experts and you got passionate opinions but very little evidence. To help lift the veil, IHME scientists had developed methods to track what affected personal health in Washington State and King County, where Seattle is located.

"There's almost a fifteen-year difference in life expectancy between the best- and worst-performing counties in the U.S.," Steve Lim said. "The King County work is about understanding in more detail what are the underlying determinants of that disparity. To what extent is it your economic purchasing power? The environment you live in? Are you less likely to be diagnosed for underlying conditions and therefore less likely to receive care? Or is it more related to the treatment of your condition once you reach a hospital?"

Individuals volunteered to be part of the study. "We know where they live—we get their nearest street intersection—so we know their access to parks," Lim explained. "We know how long it takes them to travel downtown, or to fast-food establishments, or to a supermarket." IHME also tracked the volunteers through other data systems. "What's their cholesterol level?—we have their medical records," said Lim. "If they have high blood pressure, are they taking medication?—we have their pharmacy records. If they have a heart attack on the street, how quickly do emergency services respond to them? If they're admitted to hospital, what is their treatment? If they're released, how long do they survive?" IHME had designed the study to protect privacy. The point was not to target individuals. It was to expose gaps in existing programs. Like the census, no one ever saw individual answers in the final report, only population-level statistics. "The whole idea is, what are potential policy or intervention points you can act on?" Lim said.

Still other programs were designed to work directly with patients to give them personal recommendations on health decisions they could make on their own. One of these, using a Web app co-developed with the Dartmouth Institute for Health Policy & Clinical Practice, was being tested in five hospitals in Iowa, New Hampshire, and Colorado. "If my blood pressure is 140, and I only exercise two hours a week, which should I do first: lower my blood pressure or increase my physical activity?" Lim put the kind of questions the program answered.

"We have the know-how, we have the desire, we just need to implement it," Ali Mokdad had said of his new project in Saudi Arabia, but that was true of every IHME project and every country in which they worked. "Everything ties together here," Mokdad continued. "It's like burden of disease. Having the product, Chris could say, 'See, I could do it. And see how important it is.'" King County would show the United States. Saudi Arabia would show the world.

A ll these future studies would benefit from the publicity and enthusiasm Global Burden was generating, both in print and, soon, as an online tool. They would need the goodwill, because in fact they would be much harder to finish. In five years, the same time it had taken to complete the new Global Burden of Disease research, Gakidou expected IHME's health system studies in the ABCE project to cover just fifteen or sixteen countries. "The board said to focus ninety percent of our effort on Global Burden," she said. "This is the other ten percent. And it is infinitely harder."

A new generation of health officials, comfortable with big data and unfazed by revelations of past policy missteps, was already eager to apply any lessons from the research. After one of Murray's innumerable appearances promoting Global Burden, two awed public health ministers from Yemen—admirers not critics—

approached and asked to have their picture taken with him on their phones. "Chris Murray has changed a lot of how we have seen health systems in the world," one said. "Since I was a student in 2000, it's been my dream to meet him."

The man was Jamal Thabet Nasher, the deputy minister for health planning and development in Yemen's Ministry of Public Health & Population. In 2000, he had been studying for his master's degree in public health at the London School of Hygiene & Tropical Medicine. When the 2000 *World Health Report* came out, "there was so much debate and so much controversy," Nasher said. "There were articles in international journals criticizing the ranking of the countries and they also criticized the methodology." Nasher disagreed. "It was an excellent attempt to evaluate countries," he felt. "If I am working in the ministry of health, I should be looking forward to have my health system performance get better and better. Murray and colleagues have set benchmarks for countries to improve themselves."

After Nasher joined his government, Yemen performed an internal health system assessment based on the 2000 *World Health Report* criteria of fairness, efficiency, and responsiveness. In the report, the first-ranked country overall had been France. Eighth, though, was Oman, while Yemen was 120th. "Oman is just across our borders," said Nasher. "We asked ourselves, 'Why does our neighbor rank eighth worldwide compared to other health systems?' Their economic status, their society, their geography is similar to ours. There should be something to learn. How can we be better?'"

In 2005, seventeen Yemenis from the Ministry of Health had visited their Omani colleagues on a fact-finding mission. "We learned that they are very efficient," Nasher said. "They closed some health facilities and health centers because not a lot of patients came." To compensate, other health facilities nearby extended their working hours. "In our new strategy, we do so, too," Nasher said. The next benchmark studied was responsiveness. On this front, Oman's emergency system amazed the Yemenis. "Any emergency phone call is responded to within seven to eight minutes," Nasher recounted. "That also guided us." After their visit, Yemen established its own national ambulance system to cover both the biggest cities and most-used country roads. "We have lots of mountains and highways which connect the towns and cities together," Nasher said. "This [ambulance system] has helped very much in improving our responsiveness to people."

The Yemeni Ministry of Health was still studying Oman to improve and save lives—all based, Nasher said, on the 2000 *World Health Report*. "Everyone wants to improve," he believed. "Everyone wants to be assessed by someone." Ironically, because the WHO never followed up on its health rankings, the 2000 *World Health Report* had never stopped being cited by reporters and editorialists. No single document produced by the WHO had ever attracted so much attention.* Through the rest of graduate school and into his early years in government, Nasher had wondered, "Why hasn't there been a second report? A third report? A fourth report?" When he finally met Murray in person in 2012, Nasher had just been elected chair of the program, budget, and administration committee of the WHO executive board. Come their biannual meetings, he said, "I'll be putting forward that the focus of the WHO over the coming ten to fifteen years should be based on the work that's going to be produced by IHME."

A few months before, Peter Piot, head of Nasher's alma mater, had struggled to express the present and future scope of the

---

*In the United States, Michael Moore featured it in his 2007 documentary *Sicko*, *Wall Street Journal* opinion writers attacked it anew in the debate about Obamacare, and a 2009 American viral video on YouTube, called "We're Number 37," has had 650,000 views and counting.

information that was coming from the upstart institute in Seattle. "There's nothing that comes close to the vastness of the enterprise," he said. "There's not a single institution in the United States or Europe that has IHME's resources or ability—and we are the largest school of global health in Europe." "Epic" was the word other experts kept using to describe Global Burden. Piot agreed and went further. "It's epic," he said. "The work they're doing in health systems, it's epic squared."

CHAPTER EIGHTEEN

From Galileo to Chris Murray

An investment lesson—Relationship management—"You
took the Hippocratic Oath"—The body of work—A mellow,
easygoing guy—The making of the thing.

March 2013. Seattle.

On a clear morning, you can see the Gates Foundation from the
sky as you fly into Seattle. First you spot the Space Needle,
looking like a mid-century-modern floor lamp, then the melted-
red-crayon clump of the Experience Music Project, and then the
new office buildings of the Gates Foundation, completed in 2011
and resembling a pair of boomerangs askew in relation to each
other. It is easy to imagine the yearning that first glimpse in-
spires both in locals and out-of-town visitors, particularly medical
researchers and representatives of foreign countries. *There*, they
must think, *are billions of dollars. Some of them may be for me.*

The foundation longs to do good. Others long for its good to
be done for them. Yet the fact is—and the Gateses themselves,
Bill and Melinda, are well aware of this—that all their vast fortune is a pittance compared with the costs of keeping everyone on
Earth healthy. The oft-maligned Veterans Health Administration
(VA) in the United States, serving fewer than 9 million people,
has an annual budget several times the total amount of health aid

to developing countries from all public and private sources worldwide. Every year or two, Australia spends more money on the well-being of its paltry 22 million citizens—approximately 0.3 percent of the world population—than the Gateses will be able to give away in their lifetime. The Gateses can do a lot, sure. But in a larger sense they are perhaps most valuable as an example. How to give, how to invest. A good way to do good. Not the most good. The best good with the resources available.

In supporting IHME, the Gates Foundation gave individual countries and donors tools to help maximize their own efforts. The Global Burden of Disease study, Chris Murray and Alan Lopez's meticulous decades-long creation, was now measuring the impact of 235 causes of death, 289 diseases and injuries, and 67 risk factors for men and women in 20 age groups. In March 2013, three months after the *Lancet* publication, IHME was ready to release complete country-level results, along with customizable software to help people understand it all: GBDx, now named GBD Compare, and three other interactive online visualizations to complement the previous five prepared for *The Lancet*. At last, like the Gateses, the public, the media, health professionals, and policy makers would be able to look up where they—and every human being in the world—stood. And then, with an agenda for action, they could begin making their and others' lives better.

The new release, with its accompanying press conferences, briefings, and celebrations, was being hosted by the Gates Foundation at its headquarters in Seattle. At 7:30 a.m., Tuesday, March 5, 2013, Chris Murray paced one of the foundation's large meeting rooms, dressed in a black suit. The room held seats for 170, divided into two sections. On the walls were photographs of the kinds of projects the foundation supported, ranging from the installation of broadband Internet in a Vietnamese library to the administration of vaccines to newborns in rural India. As Murray walked back and forth, Peter Speyer was at the podium, testing the Internet connection. A woman in a green dress steam-ironed the tan tablecloth on an adjacent speaker's table. In front of a satellite link and sound-mixing station a nine-person A/V team positioned three cameras.

An aide beckoned Murray. Reporters were phoning in to speak to him ahead of the scheduled speeches, panel discussions, and demonstrations. Tom Paulson, a journalist and global health blogger, would track subsequent stories that discussed the new data. News media everywhere were watching the national burden-of-disease numbers, he would find, from South Africa to Spain, India to Argentina. "Most media, Australia a notable exception, reported on how poorly [their home countries] were doing," Paulson would observe. In Great Britain, *The Guardian* would write, "Smoking, diet, alcohol and drugs are the main contributors to the UK's below average healthy life expectancy." A Chinese outlet would say, "Poor diet, smoking and pollution are leading causes of death." "Russia saw virtually no increase in life expectancy from 1990 to 2010," the *Moscow Times* would report.

For the next half hour, as people entered and took their seats, Murray sat in another, smaller room to answer questions at a call-in press conference. "One of the main purposes of Global Burden has been to make health comparable between countries," he said. "We believe these tools have the capacity to make complex information relatively simple and accessible to a broad audience. You can go cause by cause, risk factor by risk factor, and see who's doing the best and who has made the most improvement." How did the United States rank, for example? "Take something simple like life expectancy," said Murray. "There are a number of countries now doing better. If you look at the data, you see where we've had particular trouble with women. It's not all grim, though. We're doing well for stroke and breast cancer."

Could you zoom in on local differences? another reporter wondered. "It took five hundred people and five years to get to the point where we have data for one hundred and eighty-seven countries over time," Murray said. He mentioned the work in Seattle's King County, now extending to New York City and Fulton County, Georgia, where Atlanta is located. And municipalities in the United Kingdom, China, and Brazil looked to follow suit. "We want to take the Global Burden of Disease approach and apply that not only at the county level, but get down to even more local areas," he continued. "I suspect results are a year to eighteen months away."

What was the best overall indicator? someone asked. "Healthy life expectancy—how many years you can expect to live in good health," Murray said. When were they repeating the study? asked someone else. The scientist leaned forward in his seat. The big news. "The study will be updated every year," Murray said. "There will be continuous updates. There will also be the expansion of diseases and injuries and the addition of new risk factors over time. We will be looking for, and hopefully getting, partners in every country in the world."

To help make it happen, the Gates Foundation, the host of this morning's gathering, had given IHME a new grant of $25 million. "We use global burden estimates to prioritize not only work in the field, but also research and development: Where should we invest?" said Stefano Bertozzi, who led the foundation's efforts to prevent, treat, and cure HIV. "For the pharmaceutical industry, it's expected return on investment, measured in dollars. For us, it's expected return on investment, measured in improvement in global health."

After the call-in questions were finished, Murray walked back to the now-crowded meeting room, shaking hands and making or renewing contacts. Collaborators past, present, and future were here, from Coders4Africa to the UK-based Make Roads Safe, the University of Balamand in Lebanon to the New York City Department of Health and Mental Hygiene. By five minutes to 9 a.m., every seat was taken, with two to three dozen men and women standing in back. They all suddenly hushed, as if it were the moment at a wedding when everyone realizes that the bride is about to enter. "Bill is coming," went the whisper up and down the room.

Then there he was. Entering from a hidden side entrance: Bill Gates.

Gates wore black tasseled loafers, dark blue trousers, and a striped dress shirt. He peered through narrow rectangular glasses, his famous face freckled, his untidy hair light brown and gray in equal parts. In person, here at home in his own headquarters, he looked smaller, younger, and more at ease than he did in most photos taken at other places and occasions. Taking a front-row seat beside Murray, Gates smiled warmly and shook his grantee's hand.

Murray took the podium. "Good morning," he told the audience. "What I'd like to do is to use both the science behind the Global Burden of Disease 2010 study—the work of this large collaborative—and a suite of visualizations that are live on the Web right now—for any of you to use when you leave and for hopefully millions of people around the world to use in the years ahead—to tell a sort of story about what has happened in global health over the last two decades and some of the challenges ahead. It's a story of remarkable progress in many places, it's a story of an unfinished agenda, and it's a story of huge diversity in health patterns at the local level."

Without notes or slides, using only a Web browser with Global Burden visualizations projected on a screen, he displayed deaths by age and cause from 1990 to 2010. It was a newer, larger, more detailed version of the story he had told at so many meetings in

2012 and again at the study's initial publication launch with *The Lancet* in London at the end of the year. Three months later, the whole world could follow along with him. Murray moved metrics to years of life lost, lest a death at age ninety be equated to a death at age one, and then shifted views again, adding in the disabling toll of illness and injury. Each time, the visualization adjusted instantly. "What you see is this dramatic decline in child mortality, where we have reduced the number of child deaths down to, still a huge number, namely just below seven million, but still a remarkable change globally, a shift in the global death pattern towards older ages," he said. "The conditions that cause disability are quite different than the ones that are causing premature death, and when we put it all together in our metric of DALYs, which is our way of quantifying the total burden of disease, you see this complex picture for the world: a huge agenda around children, a huge volume of burden in young and middle-aged adults."

The visualizations weren't limited to what caused health loss by year. As Murray had shown with his trial run in Mexico City five months before, they also showed the numbers by age, sex, or location. Pointing and clicking, Murray lined eleven countries up against one another—"From Japan, with some of the best health in the world, or Australia, almost as good," he narrated, "right through to Niger, which probably has the highest child mortality." Between were the United States, Mexico, China, Russia, Indonesia, Guatemala, Yemen, India, Zambia, and Rwanda. He toggled again from 1990 to 2010. "Notice," Murray said, "even in the worst-off places, we've seen marked progress. But what this also shows is there are different levels of progress in different places." Health loss per person had almost halved in Rwanda in twenty years. By the same measure, China had not only caught up to but surpassed the United States. "That's a transition that's really quite dramatic," he said.

As Murray spoke, Gates took notes in black pen on the printed event agenda. He boxed key ideas, and nodded repeatedly whenever the scientist stepped through the data. With GBD Compare—formerly GBDx—Murray illustrated complex concepts like the epidemiological transition, country by country, from burden led by child mortality and infectious diseases to that dominated by noncommunicable diseases. Another tool let his audience view how different health threats ranked in order of importance within individual countries. In Zambia, showed Murray, "You have to go down to the *twelfth* cause of burden before you see something that's not a communicable, maternal, or neonatal cause." The original problems the Millennium Development Goals set out to address had not gone away, even if they were now part of a much bigger picture. Gates and Trevor Mundel, the foundation's global health head, looked from the list on the big screen to each other. They smiled sadly, nodding.

Every January, Gates released an annual letter to the public, an issue-focused, first-person take on the foundation's progress and priorities. Past themes included "innovation in agriculture," "the miracle of vaccines," and "enlightened self-interest." The topic of 2013 had been measurement. "A business has increasing profit as its primary goal," Gates wrote: "Management decides the actions—such as improving customer satisfaction or adding new product capabilities—that will drive profit and then develops a system to measure those on a regular basis. If the managers pick the wrong measures or don't do better than their competition, profit goes down."

Foundations and government programs were different, he explained. Unlike businesses, they picked their own goals and made that their bottom line. "In the United States our foundation focuses mostly on improving education, so our goals include reducing the number of kids who drop out of high school," Gates continued. "In poor countries we focus on health, agriculture, and family planning." He then added: "Given a goal, you decide on what key variable you need to change to achieve it—the same way a business picks objectives for inside the company like customer satisfaction—and develop a plan for change and a way of measuring the change. You use the measurement as feedback to make adjustments. I think a lot of efforts fail because they don't focus on the right measure or they don't invest enough in doing it accurately."

When Gates took the podium after Murray, he elaborated on the importance of accurate and detailed data. "In almost every endeavor, but particularly in health, it's the areas where we go in and do a good job of measurement that we make progress," he said. "We see who's doing well, we see who's not doing well, and we come up with the tactics to make very rapid change."

He talked about reading the first global burden of disease figures in the 1993 *World Development Report.* "I was completely stunned by the burden of disease in poor countries, to see that diarrhea was killing literally millions of children, and that some of those causes of diarrhea, like rotavirus, were preventable—that is, there was a vaccine that was available in rich countries, but ironically not in poor countries, that could bring those numbers down." Having been shown the data, he had decided to show the money: since 1999, the Gates Foundation had committed $2.5 billion to expand vaccine access and development—a major reason 2.5 million fewer children had died in 2010 than in 2000. "It was seeing that data, that early visualization that's nowhere near what we've got today, that got the Gates Foundation on the track of focusing on global health."

Today any country in the world could join them to improve the lives of its people—or people anywhere. "As we've been going down that path, we've had a chance to fund a lot of studies to go out and measure things, but there's never been anything that could pull the data together, to be the sort of ultimate communication tool for the various debates in the field and the various policy decisions to be made. And now with the Global Burden of Disease, we have those tools," Gates said. "I want to congratulate Chris Murray and his team for having done a phenomenal job."

In 2012, an independent external evaluation team jointly hired by the University of Washington and the Gates Foundation had given IHME an "A" for technical excellence and a "C" for external cooperation. In terms of key achievements, the evaluators reported, the institute's work was considered "state of the art." Its role of challenging the status quo was beneficial to the field. Murray's team was "highly credible," "highly independent," and had already made a "dramatic impact." In terms of major misses, however, IHME's first five years had resulted in an unhealthy "alienation" of others in the health metrics field. Partnership building suffered because of the perception that "IHME unnecessarily provokes the UN system and creates tension." As the IHME board member and Public Health Foundation of India president K. Srinath Reddy put it, "Clearly, IHME is widely respected, but not universally loved." Richard Horton offered a more upbeat spin. "Good science is polarizing," he said. "From Galileo to Chris Murray."

A year later, better cooperation was very much part of the ongoing agenda. The same day national Global Burden results were being announced at the Gates Foundation, *The Lancet* published a detailed analysis specific to the United Kingdom. Murray, Lopez, and other core scientists from the larger Global Burden of Disease study were listed authors. So were local leaders at the UK Department of Health and Public Health England, the National Cancer Action Team, the King's College London Dental

Institute, and the Vision and Eye Research Unit at Anglia Ruskin University in Cambridge. It was the first country-specific report done using the newest Global Burden data, and the impact was as much a model of the new era of public health agenda-setting as the report itself.

Jeremy Hunt, the British secretary of state for health, responded instantly. "I want us to be up there with the best in Europe when it comes to tackling the leading causes of early death, starting with the five big killer diseases—cancer, heart, stroke, respiratory and liver disease," he wrote. "But the striking picture of our health outcomes across these major causes of early death published in *The Lancet* recently shows that we have a long way to go before we are confident that we can achieve this aspiration." A set of specific policy proposals followed, entitled "Living Well for Longer: A Call to Action to Reduce Avoidable Premature Mortality." To Murray, the British experience was an important test of working with local collaborators. "Having local ownership of the study—somebody who actually understands what it's about, knowing all the numbers, with all their pros and cons, and how they're useful—has been very impressive to me," he said afterward. "The way forward is to be more aggressive about engaging individual countries."

To complete Global Burden, Murray had recruited some five hundred outside experts, organized globally by their disease, injury, and risk factor specialty, as colleagues. Now he wanted to establish another equally large panel of authorities who were experts on the health conditions in individual countries. "So there is somebody," he said, for example, "who is the expert on cause of death or noncommunicable diseases in Kenya." To that end, the new Global Burden team would have permanent regional directors recruiting collaborators and building up the capacity for local burden studies worldwide. "A whole part of this is relationship management," Murray said, "so that's the adventure we're now embarked on."

Ali Mokdad was already signing up nations in the Middle East. Rafael Lozano, soon to split time between IHME and the National Institute of Public Health of Mexico, would take Latin America and the Caribbean. Alan Lopez was going to link Australia to the Pacific and parts of Southeast Asia; Heidi Larson, a medical anthropologist and senior lecturer at the London School of Hygiene & Tropical Medicine, was finding European partners; and Tom Achoki, a Kenyan doctor, public health specialist, and former IHME fellow, would cover African initiatives from a base in Botswana. Already, in Rwanda, after Global Burden findings indicated that household air pollution was the nation's leading risk factor, government leaders had announced a new program to install more than a million new clean cookstoves. "I think we're entering the brave new era where burden of disease becomes mainstream for most countries," Murray said.

Even the factious U.S. health care system was uniting to understand the new numbers. On July 10, 2013, the *Journal of the American Medical Association* published "The State of US Health, 1990–2010," co-authored by Murray's team and collaborators from more than fifty other American medical and public health institutions. Measuring the burden of diseases, injuries, and leading risk factors in the United States, and comparing those measurements with those of the thirty-four countries in the Organisation for Economic Co-operation and Development, they found that in 2010 the United States ranked 27th in terms of age-standardized death rates, 27th in life expectancy at birth, and 26th in healthy life expectancy. The country's top risk factors related to burden of disease were dietary risks, smoking, high body mass index, high blood pressure, high blood sugar, and physical inactivity. More than 678,000 deaths in 2010 were attributable to poor diet alone.

The morning of the publication, Murray presented these findings at a White House event for mayors and other local officials hosted by First Lady Michelle Obama as part of her Let's Move! public health initiative. An accompanying interactive online U.S. health map gave county-by-county assessments of life expectancy, physical activity, obesity, and blood pressure. Depending on the county, IHME leaders observed, men and women might have the same average life expectancy in Indiana and Panama, Nevada and Vietnam, Michigan and Syria. This was kind of the comparison politicians would find very difficult to ignore.

In November 2013, less than a year after the first Global Burden publications in *The Lancet*, the WHO released its own provisional global burden-of-disease estimates for the years 2000–2011, "consistent with and incorporating UN agency, interagency and WHO estimates for population, births, all-cause deaths and specific causes of death." Yet these were only Excel spreadsheet files, only for continent-wide regions, not individual countries. How would they catch up? "The role of the WHO in my view should be not to replicate what academic institutions can do much better," Lopez said. "Murray and I have an army of people working with us—we'll get there [to completed, in-depth burden of disease studies] much faster."

After ten years at the University of Queensland, Lopez had moved to the University of Melbourne and established a new Global Burden of Disease subgroup. A primary focus would be improving countries' vital registration systems. Maybe being counted when you are born and die is not widely considered a fundamental human right, he acknowledged. But without it every other right—from food and shelter to education and the ability to vote—is at risk, and essential improvements in personal and public health may be impossible. "There are about fifty-two million deaths a year, of which about thirty-five percent—seventeen or eighteen million—are recorded," Lopez noted tersely. "So we're missing sixty-five percent of deaths." Still. "I carry these numbers in my head everywhere I go," he admitted.

IHME was changing, too. Between January 2012 and June 2014, the number of people working there almost doubled, to close to two hundred people, among them twelve data indexers, twenty-six faculty members, thirty-five fellows, and forty-four researchers. The institute started a Ph.D. program in health metrics and evaluation at the University of Washington and occupied an additional floor in its building. Other funding sources—scientific and philanthropic grants, government and aid group contracts—were starting to balance out the vast contributions of the Gates Foundation. IHME launched an annual $100,000 prize to recognize "individuals or groups that have used Global Burden of Disease data to take action that makes people healthier." (Funding came from David Roux, an IHME board member and cofounder of the private equity firm Silver Lake, and his wife, Barbara.) Academic association notwithstanding, the atmosphere was prosperous biotech firm or successful dot-com. One day Steve Lim read an e-mail from a departing staffer, saying goodbye to everyone on her final day. The group had grown so large that he had never met her.

"IHME is a juggernaut," said an outside collaborator. If it rather than other organizations received most of the available funding for health measurement and evaluation, he worried what would happen to the next generation of high-level global health researchers. "Global Burden could excite young scholars and bring them in," he said. "Conceivably, it could also crowd them out."

This was a concern Murray shared. "When there's just one monolithic source, mistakes can be perpetuated," he said. "They're rolled over for generation after generation. Competition is a safety on being wrong." The problem was, about the only areas in which other researchers were doing the same kind of work was in tracking the Millennium Development Goals—child and maternal mortality, TB, malaria, and HIV. In comparably measuring almost everything else that hurt people, more than two thirds of the global burden of disease, IHME stood alone. "It's not great," Murray admitted. But at some point, it was up to the rest of the world to catch on. Murray didn't just want to replace old authorities. He wanted to spread his new ways of thinking—and his fervor to turn that thinking into specific plans of action. Given the way the latest findings had been received, he had reason to be hopeful.

Murray and Gakidou hosted a party at their home on the evening of the Gates Foundation event. Murray himself, accompanied by Alan Lopez, arrived only minutes before the party was supposed to start. He opened the door and Kuma, his ninety-pound yellow lab ("Kuma" means "bear" in Japanese), pounced. Murray hurried the dog to a closed-off area as the doorbell rang. "Emm," Murray called to his wife as he looked out the window. "Our first guests are the Gateses."

Not the former head of Microsoft now turned philanthropist, but his father. A big man, well over six feet tall, Bill Gates Sr. wore slacks, a sports jacket over a brown sweater vest, and a hearing aid. His glasses were the same style as his son's, but thicker-framed. He wore them slid down toward the end of his nose. With him was his wife, Mimi, a small-framed and kindly mannered woman, dressed in black pants, a black shirt, and a silver necklace. A nanny helped eighteen-month-old Natasha Murray down the stairs to help greet the guests. The toddler's hair and eyes were dark like her mother's, but her grin was all Murray. "Do you want Mama or Poppa?" Gakidou asked her in

Greek. "Mama!" she shrieked. Murray took her hand anyway. Natasha laughed uproariously. Soon she was leading him around the cleared living room floor.

The house was new, large but not enormous, on a corner lot in Seattle's Magnolia neighborhood. (In 2012, Murray's salary was $488,000—less than a University of Washington assistant football coach and about the average for a cardiothoracic surgeon in the United States.) From the street, you could see Puget Sound. In preparation for the party, the living room couches had been pushed against one long room wall. Mimi Gates knelt to meet the toddler. "Hi," she said with grandmotherly practice. Gates Sr. took a high stool by the open kitchen and talked to the nanny. Most weeknights she left at 5:30 p.m. She didn't work weekends. Murray liked that. "I'm not in med school," he said. "I can spend time with my daughter." He had recently solo parented for a week while Gakidou traveled. "It was good," said Murray. "She's so much fun. The thing about being an older parent is that all the things that stress you out as a younger parent, age makes vastly more enjoyable."

The party swelled. A bartender served wine and grapefruit-infused vodka. Soon several leading figures in global health, including the host, were loosening up. "The one thing that gets me emotional is my parents or my family," Murray had reflected in an earlier conversation. Small matched paintings on the living room wall depicted snowcapped mountains at sunrise. They were reminders of what might have been a personal tragedy. Murray and Gakidou had celebrated her thirty-fourth birthday in March 2008 by renting snowcats and skiing backcountry central British Columbia with guides. "We've done a lot of crazy stuff," Gakidou said. "But the time I got caught in an avalanche was the one time we went skiing with a guide." Murray had gone first down the gladed slope. Gakidou, following, was swept down and buried.

She crushed two vertebrae and broke her neck, a hand, two ribs, her femur, and every bone between her knee and ankle. Evacuated by helicopter to Calgary, and then moved again to Seattle, she underwent nine surgeries and had to spend close to four months recovering in hospitals or in bed.

"Emm's nine surgeries, almost dying—we thought for a while she'd be paralyzed—it gives you a whole new perspective on the care-giving element of health systems," Murray said. He remembered waiting in the first hospital to which she had been taken. Her doctor paged Murray. When they met, it turned out that he wanted Murray's credit card number. He wanted to make sure he'd get paid for his surgery because they weren't Canadians. "I stared at him," said Murray. "He said, 'You think this is inappropriate?' I said, 'You took the Hippocratic oath.'"

Perspective mattered. "It's harder when it's someone close to you than yourself," said Murray. "It's why we'd better be right. All of this stuff is too important. The first lesson of medicine is do no harm." When you worked at the level of global or national health policy, mistakes could kill millions of people. Discoveries could save that many and more. "That's why we're so obsessive," Murray said. "It's also why a competitive world in this space is a good thing." A generation ago products like Google Earth were science fiction. Now we used them every day. IHME promised the same kind of perspective for health. Study it and we could all find routes to living better.

Soon it was time for Natasha to go to bed. "What I'd like is to find a way for her to understand that there's a bigger world out there," Murray had said earlier. "That there are things she should see and understand. Kids start with this tiny little worldview, which is them and their parents, and it slowly broadens. Left to their own designs, people's worldview is really insular. So how do you counteract that?"

Certainly no one who was employed at IHME could be accused of a narrow view of the world. Not everybody was staying on the Global Burden team after the country-results release, but few who ever worked with Chris Murray, or simply existed alongside him, were untouched. Catherine Michaud, from the Harvard Pop Center and HIGH, the only person besides Murray and Lopez involved in every Global Burden of Disease study for two decades, considered herself retired after the Gates Foundation event. "Nobody knew at the beginning that Global Burden would survive," she remembered. It had done so because "Chris and Alan are flexible," she said. "Some of the basic assumptions that went into the DALY metric have been completely revised." Global Burden lasted because it was a work in progress. It always would be improving its consideration of old numbers while adding new ones. Michaud flew to Seattle for twenty-four hours, hugged Murray, and said congratulations. "I'm glad it's over," he told her. "What's next?" Michaud said. Murray didn't even pause long enough for another breath. "The next version is what's next," he said.

Challenging as it might be to work with Chris Murray, it was much more productive to be his collaborator than his competitor. You were on your toes in his company. You spoke faster. You thought sharper. You had to be clear. You had to be able to explain and defend your analysis. You had to show why you were right—or at least trying to be. "Chris is not one who's going to lose sleep over style points, but look at the body of work," said his medical school classmate Jim Kim. "Think about how much we know now because of the Global Burden of Disease."

On July 1, 2012, in a wonderful twist, Kim, the activist and aid worker, had taken office as the twelfth president of the World Bank. He and Murray hadn't lived in the same city since Murray left Harvard in 2007. But when they saw each other, they immediately picked up where they had left off. "We catch up really quickly," Kim said. "The intensity with which we share a mission, which is to tackle poverty, tackle poor health, relentlessly go after improving health and well-being for the world's poorest people, we get right into business." No one who was really sick wanted to go to a doctor satisfied with the status quo. With the ongoing Global Burden of Disease study, Murray's patients now numbered 7 billion people. "If I want to know what the hell's happening in the world around a particular issue," Kim continued, "the first person I call is Chris."

Murray's commitment to his work, and his fellow workers, was intense. Those who couldn't stand his style left for other places. And those who weren't involved at all could become very estranged indeed, especially if they were family. Following more than fifteen years of litigation, he had lost all visitation rights to his three children from his first marriage. Murray still talked to them on the phone, however, and he and they e-mailed each other. Two were in the French university system. The youngest was in high school. "I tend to suppress a lot of unhappy memories," he said. But losing direct contact had to have been singularly searing. "I think a lot of this impersonal stuff is all protection," said Richard Horton. "The outer exterior of Chris you see is to stop him from being hurt."

Did his marriage to Gakidou and the birth of their daughter, Natasha, represent a second chance at family, the same way Gates had given him a second chance at a truly comprehensive understanding of the burden of disease? Late one afternoon in his IHME office, Murray blanched at the comparison without necessarily disagreeing. The greater part of his charisma, such it was, came from *not* caring what others thought. The work said everything. As a boss, this made his praise so valued; it was as close as you came to an objective assessment of your own scientific achievements. But being a father was a life, not a job. "I was a pretty driven early-thirties parent," he admitted. His brow wrinkled and then he broke into a smile. "I'm now a more mellow, easygoing guy," he observed, with an understanding that, whatever his personal progress, "mellow" and "easygoing" were not the first adjectives others would choose to describe him.

A story: In June 2012, the week before the IHME board meeting, Murray and Alan Lopez had met in Washington, D.C., to complete their core papers before submission to *The Lancet*. They had four days to finish what was likely to be among the most-cited health research of all time. Still, Murray, while refusing all other breaks, kept insisting that they visit any organization that could help add to or make use of Global Burden findings. In a single day, they went to PAHO, the Pan American Health Organization, regional office of the WHO for the Americas, where the director, Mirta Roses Periago, said she would convene a study of anemia in the region; to USAID, the United States Agency for International Development, where he brokered a deal to improve data collection and cut time to publication by eighteen months; and to a meeting with U.S. Centers for Disease Control and Prevention researchers, in from Atlanta, who asked his help in seeing where their models of recent flu pandemics had gone wrong.

Not even the president of the United States could stop Murray's movements. Three times, when blocked by a passing White House motorcade, he hopped out of his taxi and walked. Lopez worried that they'd get lost. Impossible, Murray said. His sense of direction was infallible. "It's the same reason when I was a kid, and we were crossing the Sahara," he said, "my father had me do the navigation."

He was no longer a young boy, but he would always be a member of the public health initiative that was Murray, Murray, Murray, and Murray. You could say the years in Africa changed Chris. You could also say it showed him who he really was. Part of a brilliant, stubborn, even foolhardy family—at once selfless and determined, sophisticated and repressed, intellectual and almost crazily daring. And, ultimately, impossible to stop.

Linda Murray, Chris's oldest sister, had retired from her job as a flight attendant for Pan Am before he graduated from college. She raised two children with her husband and they now lived part-time in Oregon and part-time in southern California.

In the 1990s, in collaboration with UN peacekeepers, Nigel Murray had helped reorganize Bosnia's central health system. For his service, he was named a member of the Order of the British Empire. Now, following executive stints in health care administration in New Zealand, he had moved with his wife and children to British Columbia, Canada, where he oversaw a $2.9 billion, 22,000-employee, 1.6-million-patient health and hospital network, Fraser Health. "I realized I could be more influential as a manager of medical care for a population," he said, "influencing the system that delivers it."

Megan Murray had become a professor of epidemiology at the Harvard School of Public Health and director of research at Partners in Health, the nonprofit organization Paul Farmer and Jim Kim co-founded, where she tracked multiple-drug-resistant tuberculosis and aided in implementing effective local-led interventions around the world. "Providing direct medical care can be done with government support," she said, not "one little family." Some regarded her as even more intense than Chris. On a recent work trip to Rwanda, she had detoured to Nairobi with her husband and children and taken one of the original Land Rovers out of storage to show them. It was still where her parents had last parked it. And, with a little coaxing, it still ran. She remembered her father running out of penicillin. "We were there," she said. "We couldn't provide care without a supply chain." Partners in Health wouldn't open a new hospital without the certainty of supplies. "That comes out of the experience of watching people die without medication," said Megan. "That's why we all got into public health."

Chris agreed. "It made career choice pretty easy," he said about his childhood experiences. "I really didn't think about it." When entering a foreign country and asked to list his occupation, he still wrote physician, even if his "practice" now consisted not of seeing individual patients but tracking the health of billions. The beauty of medicine, said Murray, was "There's a lot you can fix." Overall, the average burden of disease per person had been cut almost 25 percent between 1990 and 2010. "If we can make all this analytic stuff work and people can learn lessons," he said, "we can do that again sooner than 2030." If we chose to, people everywhere could focus on what was really hurting us—and then how best to heal.

John and Anne Murray kept returning to Africa as primary health providers for a full two decades after their children had set out on their own. "The rest of us kids went off to college," Chris said. "They kept going." Four months a year, from 1980 until 1995, they went to Kenya. Then, for three years after John retired from the University of Minnesota, they moved to Malawi, where he taught medicine in addition to his doctoring until 1998. "We were getting a bit old at that time," he observed. Both were seventy-seven when they quit.

Anne died of a stroke in late April 2009 at the age of eighty-eight. John moved back to New Zealand. A severe hip problem limited his movement, but he chose to live in a cottage on the farm where his wife grew up, in sight of where she was buried. "It's very sentimental for me here," he said. "I can sit with nobody bothering me and look out at the dairy cows." For his ninetieth birthday, he joined his scattered family for a celebration in Seattle. Over a three-hour meal, everyone offered toasts, John included. Visibly moved, he gave thanks for his marriage and children, saying how grateful he was for the privilege of being part of a family that loved each other so much.

A little more than a year after this dinner, driving toward the same restaurant where the meal had been, Murray was asked what the fun part was for him—making the Global Burden of Disease study or presenting it to the world? He didn't hesitate. "The fun part was the making of the thing," he said. He was approaching the age his parents had been when they took him and his siblings to Diffa. For all his travels over the past forty years, he had never been back.

Someday.

"I'd love to go," Murray said. "I loved living in Africa as a kid. I still like dry hot places. They remind me of the desert."

He had been to many other places, though, and throughout he was still the navigator. The one constant in all his journeys was that he was always trying to reach the place where the weight that disease and injury placed on humankind was as unburdensome as we could make it. He could appreciate how far the world had traveled since 1973. And that we still have a long way to go.

How to Live a Longer and Healthier Life According to the Global Burden of Disease

The beauty of the Global Burden of Disease approach to understanding health is that it can be used in so many ways. Governments can use the study and related research to chart policy, health departments can use the data to help allocate resources, and individuals—you and I—can use it to better take control of our own future. If you want to live the longest and healthiest life possible, here are eight ways how.

1. BEAT THE REAPER.

There's no escaping death, but if you know the enemy you can take preventative steps to ensure the longest possible life span. Begin by consulting the interactive visualizations available from the Institute for Health Metrics and Evaluation. In the diagram below, for example, Global Burden data point to five causes—ischemic heart disease (IHD), lung cancer, stroke, chronic obstructive pulmonary disease (COPD), and diabetes—that accounted for more than a third of years of life lost to early death in the United States in 2010.*

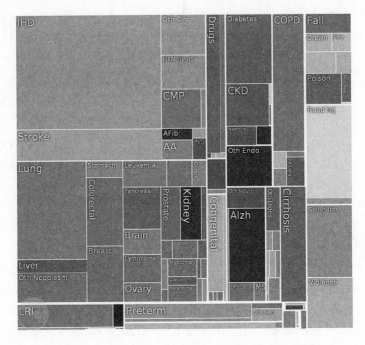

Causes of Years of Life Lost to Early Death in the United States (2010)

What does this mean for you personally? For starters, you could eat better: if the average American pursued the ideal diet identified

---

*AA: aortic aneurysm. AFib: atrial fibrillation and flutter. Alzh: Alzheimer's disease. CKD: chronic kidney disease. CMP: cardiomyopathy and myocarditis. COPD: chronic obstructive pulmonary disease. HTN Heart: hypertensive heart disease. IHD: ischemic heart disease. Int Lung: interstitial lung disease and pulmonary sarcoidosis. LRI: lower respiratory infections. Med Treat: adverse effects of medical treatment. Oth Circ: other cardiovascular and circulatory diseases. Oth Diges: other digestive diseases. Oth Endo: other endocrine, nutritional, blood, and immune disorders. Oth Neo: other neonatal disorders. Oth Neoplasm: other cancers. Oth Neuro: other neurological disorders. Parkins: Parkinson's disease. PVD: peripheral vascular disease. Road Inj: road injury.

by IHME researchers (see more in item #3), years of life lost to heart disease would fall by 87 percent and years of life lost to stroke would fall by 67 percent, according to Global Burden estimates. Increasing physical activity or lowering body weight, meanwhile, could cut the toll from diabetes 30 to 75 percent. Put out cigarettes completely and more than 80 percent of the health loss from both lung cancer and COPD would be eliminated. If you're a smoker, this means you.

2. STAY STRONG TO THE END.

In the United States, six of the top ten causes of disability kill no one directly. These are low back pain and neck pain, osteoarthritis and other musculoskeletal disorders, and major depressive disorder and anxiety disorders. What's more, the leading sources of pain and suffering are remarkably similar worldwide. If you suffer from any kind of recurrent disabling condition, you know how this limits your life. Take efforts to address your nonfatal health problems and you will gain an average of up to forty days of healthy life annually, according to Global Burden.

To prevent back and neck pain, take regular stretching breaks at work, exercise with a focus on core strength, and consult guides to improve your posture. For osteoarthritis and many other musculoskeletal disorders, a combination of medication, therapy, and surgical procedures can often relieve pain and increase freedom of movement. Depression, anxiety, and drug use disorders can be treated or cured with a range of proven interventions, including therapy. These methods work and you should be able to afford them: the 2010 U.S. Affordable Care Act, also known as Obamacare, requires health insurance plans to offer mental health and substance use disorder services, including counseling.

Top 10 Causes of Years Lived with Disability (2010)

GLOBAL	UNITED STATES
1. Low back pain (up 43%)	1. Low back pain (up 25%)
2. Major depressive disorder (up 37%)	2. Major depressive disorder (up 43%)
3. Iron-deciency anemia (down 1%)	3. Other musculoskeletal disorders (up 28%)
4. Neck pain (up 41%)	4. Neck pain (up 29%)
5. Chronic obstructive pulmonary disease (up 46%)	5. Anxiety disorders (up 21%)
6. Other musculoskeletal disorders (up 45%)	6. Chronic obstructive pulmonary disease (up 34%)
7. Anxiety disorders (up 37%)	7. Drug use disorders (up 30%)
8. Migraine (up 40%)	8. Diabetes (up 56%)
9. Diabetes (up 67%)	9. Osteoarthritis (up 56%)
10. Falls (up 46%)	10. Asthma (up 21%)

---

3. CHANGE WHAT RISKS YOU CAN.

Relatively few risk factors account for a large proportion of the global burden of disease and, in the United States as worldwide, dietary risks in aggregate topped the list. The good news is that these may also be the easiest for you to alter to powerful effect. Surprisingly, for most Americans the best way to improve one's diet is to consume the optimal amount of fruit and nuts and seeds—300 grams daily of fruits, 114 grams weekly of nut and seed foods, according to Global Burden. Cook yourself, if possible, so you can control the amount of salt in your meals, and avoid processed meat—bacon, salami, sausages, and deli ham, turkey, and pastrami—

to prevent colon and rectum cancers, diabetes, and ischemic heart disease. If you like seafood, great—getting enough omega-3 fatty acids is as important as whole grains. If not, low-cost supplements are available to get the recommended 250 milligrams a day.

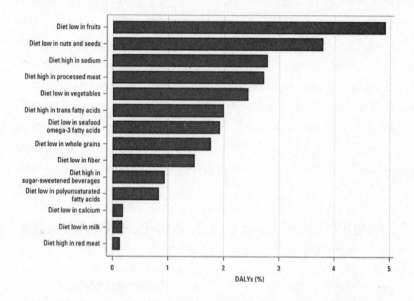

Dietary Risk Factors Ranked by Attributable Burden of Disease in the United States (2010)

Even new dietary choices lower in the rankings are well worth your consideration. According to Global Burden, if Americans had cut out drinking sugar-sweetened beverages in 2010, the overall health gain would have been more than three times greater than that from completely clearing all secondhand smoke.

4. SUPPORT PUBLIC ACTION.

When your individual efforts are joined with an effective public health campaign, you and millions of others can make rapid gains in life span together, as IHME's analysis of U.S. counties shows. Three of the five U.S. counties with the greatest increases in life expectancy between 1985 and 2010 were in New York City, which has led the nation in HIV prevention efforts and services, the creation of smoke-free public spaces and increased cigarette taxes to discourage consumption, restaurant menu calorie counts and a trans fat ingredient ban, and, most recently, an action plan to eliminate traffic deaths entirely by 2024. The end result has been extraordinary: men in Manhattan, Brooklyn, and Queens can expect to live almost a decade longer than they did a generation earlier. If all Americans could match the life expectancy improvements of those in New York and other leading counties, the United States would have the highest life expectancy of any country in the world.

Top 10 U.S. Counties in Terms of Change in Life Expectancy (1985–2010)

MALE LIFE EXPECTANCY IMPROVEMENT	FEMALE LIFE EXPECTANCY IMPROVEMENT
1. New York, New York (13 years)	1. New York, New York (8.4 years)
2. San Francisco, California (10.6 years)	2. Loudon, Virginia (7.8 years)
3. Kings, New York (9.8 years)	3. Kings, New York (6.7 years)
4. Loudon, Virginia (9.6 years)	4. Bronx, New York (6.4 years)
5. Bronx, New York (9.6 years)	5. Gunnison, Colorado (6.3 years)
6. Washington, D.C. (9.4 years)	6. Pitkin, Colorado (6.3 years)
7. Forsyth, Georgia (9.2 years)	7. Marin, California (6.3 years)
8. Goochland, Virginia (9.2 years)	8. Prince William, Virginia (6.1 years)
9. Alexandria, Virginia (8.8 years)	9. San Francisco, California (6.1 years)
10. Hudson, New Jersey (8.6 years)	10. Beaufort, South Carolina (6 years)

----

5. GET SMART (EVEN IF YOU CAN'T GET RICH).

Most economists believe that if the average wealth of a country increases, its populace will be able to afford better health care, and become healthier as a result. However, IHME data shows that this is true only up to a certain point. A much stronger correlation exists between health and education. Mexico, for instance, has a fifth the gross domestic product (GDP) per capita of the United States, but, for women, more than 50 percent of the latter's schooling. In line with the trend, Mexico's female adult mortality rate is only narrowly higher. Vietnam and Yemen have roughly equivalent GDP per capita. Yet Vietnamese women average 6.3 more years in school and are half as likely to die between the ages of fifteen and sixty. Outliers are Switzerland and Zimbabwe, with one of the world's lowest and highest female adult mortality rates, respectively, and a 200-fold difference in GDP per capita.

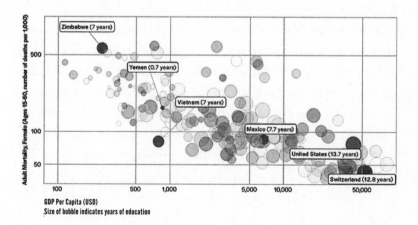

Avoiding Early Mortality—Education Improves the Odds

Educating women is an especially wise health investment on two fronts: first, it makes them better advocates and decision makers for themselves and their families during times of medical need; second, it leads women to delay the onset of motherhood and have fewer life-risking pregnancies overall.

6. BE BORN IN JAPAN— OR SWITZERLAND, OR SINGAPORE, OR . . .

Okay, you can't really control this, can you? But it's illuminating to know where your country stands compared with others in terms of individual health, and then instructive to consider how to emulate the best. According to Global Burden, average healthy life expectancy at birth—how many years you can expect to live in good health—was approximately fifty-eight years for men and sixty-two years for women in 2010, a decade less than life expectancy alone. Individuals in some places had it better than those in others, however, led by Japan.

One theory explains Japan's success as a two-part story. First, after World War II, when health improvements took off, the Japanese were already highly educated and hygienic in an egalitarian society with strong government-led public health programs, particularly for the control of tuberculosis. This led to low rates of leading communicable diseases. Second, in more recent decades, traditional positive patterns of diet and physical activity in Japan were bolstered by new public health programs to reduce salt consumption and new universal primary care programs for high blood pressure. This led to low rates of leading noncommunicable diseases.

Americans could benefit greatly from a similar focus on eating well, being physically active, addressing leading causes of death and disability, and improving access to health care for all. If men and women in the United States had the same outcomes as the Japanese, they would gain an average of four years of healthy life.

TOP 10 COUNTRIES FOR HEALTHY LIFE EXPECTANCY AT BIRTH (2010)

MALE HEALTHY LIFE EXPECTANCY	FEMALE HEALTHY LIFE EXPECTANCY
1. Japan (68.8 years)	1. Japan (71.7 years)
2. Singapore (68.1 years)	2. South Korea (70.3 years)
3. Switzerland (67.5 years)	3. Spain (70.1 years)
4. Spain (67.3 years)	4. Singapore (70 years)
5. Italy (66.9 years)	5. Taiwan (69.6 years)
6. Australia (66.8 years)	6. Switzerland (69.5 years)
7. Canada (66.7 years)	7. Andorra (69.3 years)
8. Andorra (66.7 years)	8. Italy (69.1 years)
9. Israel (66.7 years)	9. Australia (69 years)
10. South Korea (66.7 years)	10. France (68.8 years)

---

7. GET SICK IN FRANCE—
OR ITALY, OR SAN MARINO, OR . . .

Again, this is not really under your control. But it's one way to encourage your political leaders to do better. The 2000 *World Health Report* measured health system performance by country. Of the ten nations that spent the most per capita on health in 2000, only two had health systems that were found to be among the top ten performers. In terms of overall performance, for example, the U.S. health system was ranked 37th in 2000, despite the highest per capita costs. The French, ranked first in the world for their health system, had significantly lower rates of long-term disability and early death than Americans while spending less than half as much per person. In part because of the controversy these rankings produced, they have never been repeated—but that doesn't mean that they were wrong.

PERFORMANCE VS. SPENDING (2000)

TOP 10 COUNTRIES FOR OVERALL HEALTH SYSTEM PERFORMANCE	TOP 10 COUNTRIES IN TERMS OF HEALTH SPENDING PER CAPITA
1. France	1. United States
2. Italy	2. Switzerland
3. San Marino	3. Luxembourg
4. Andorra	4. Norway
5. Malta	5. Iceland
6. Singapore	6. Japan
7. Spain	7. Monaco
8. Oman	8. Denmark
9. Austria	9. Austria
10. Japan	10. Germany

---

8. STAY TUNED.

The Institute for Health Metrics and Evaluation hosts the ongoing Global Burden of Disease study and other research, including publications, policy reports, country profiles, and data visualizations, online at *www.healthdata.org.*

The core interactive visualization tool—GBD Compare—uses maps, charts, and diagrams to compare health levels, trends, and the causes of death and disability within countries and around the world. For everyone who loves lists, the GBD Arrow Diagram tool offers easy explorations of disease, injury, and risk factor rankings by region, country, age group, sex, and time period. Other tools present different ways to view the same information, and explore health challenges and successes by country, sources of data and their distinctions, and healthy life expectancy versus life expectancy, among many other features. These and more interactive visualizations—from a county-level U.S. health map to graphs of international aid dollars for health—are available via the IHME data visualization homepage: *vizhub.healthdata.org.*

A curated and annotated collection of many of the above offerings, and further stories and visuals inspired by *Epic Measures,* is available at *epicdemiology.com.* Learn more and make contact there or at *jeremynsmith.com.*

As a writer, I specialize in profiles—of people, and of ideas. I'm pretty good at math and able to follow fast-paced conversations, and I like to think of myself as an energetic guy. Following Chris Murray and his colleagues stretched my mind, exhausted my body, and inspired me with the constant example of how exhilarating hard work can be. To all the people whose stories are in *Epic Measures*, I am profoundly grateful. To all the many others who contributed so much without being named, I owe the greater part of my insights on the many topics contained herein.

Abie Flaxman first invited me inside the world of the Institute for Health Metrics and Evaluation. Little did he realize that a morning's guided tour would extend for three years and counting. Once I was in it for the long haul, Bill Heisel, IHME's director of communications, and Linda Ettinger, Chris Murray's executive assistant, worked with me on a weekly—and sometimes minute-by-minute—basis. Bill helped me reach busy people at their busiest, made time to answer pressing questions at any day and hour, and encouraged me—as did everyone at IHME—to tell the fullest possible story without regard to any party line. Linda helped me follow Chris, or at least try to keep track of his comings and goings—a more-than-full-time job. That Chris and Alan Lopez agreed to let me observe them and their global team without restrictions or control was an extraordinary gift.

I began this project as an article for *Discover* magazine, and it was a pleasure working with my editor there, Pam Weintraub, and my fact checker, Fangfei Shen. I would not have proposed writing this book without the advice of Mark Sundeen. My agent,

Michelle Tessler, responded immediately to the idea, as did my editors, Karen Rinaldi and Jake Zebede. Michelle, Karen, and Jake are smarter and more capable than I am, but they trusted me anyway to tell the right story the right way—and then saved me whenever I went wrong. For their support I cannot express enough gratitude.

Throughout, I benefited greatly from feedback on drafts in progress from Fred Haefele, Kristin King-Ries, Larry Mansch, Haley McMullan, and Mary Jane Nealon. My mother, Jane Smith, is not only the best writer I know, but a brilliant editor, and her many essential suggestions made this book better in every draft. I am blessed beyond measure for the support and good humor of my entire family—my father, Carl; my sister, Lucia; my wife, Crissie; and my daughter, Rasa. Crissie, in particular, is the partner who always said yes when I asked if I should wander the world, and who always made me feel at home the instant I returned. She makes the impossible not just possible but necessary, too.

In New York City, I stayed with Josh Engelman and Eve Teipel. In Washington, D.C., I stayed with Greg and Margot Squires. In Massachusetts, I stayed with David and Camille Bernstein, Kevin Moore and Felicity Aulino, and Josh Schanker. In Seattle, I stayed—seven times—with Ping Xu. All offered not only their friendship and hospitality, but wisdom from personal and professional experience. I shamelessly stole as many of their good ideas as I could to put in print.

Other friends whose encouragement and advice sustained me during the writing process include Tod Bachman, Amanda Dawsey, Nick DeCesare, Matthew Frank, Max Lieblich, Sam Mills, Kisha Schlegel, Rob Schlegel, Sharma Shields, June Spector, Kate Stetsko, Amie Thurber, Heidi Wallace, and Jason Wiener.

This book is dedicated in part to my high school math teacher,

John Benson, and to the members of my high school chemistry and physics study group. He and they taught me to see, understand, appreciate, and share the stories numbers tell. Doing so has been one of the great privileges of my life, and, now, one of the great efforts. I am happy to say that I loved every minute of it. Thank you all.

This book is based primarily on a dozen reporting trips and more than a hundred interviews. It would be impossible to list everyone who spoke with or assisted me in my research, but I would like to express my gratitude, incomplete as it is, to the following individuals and institutions. Everything I wrote was improved by their contributions. Of course, I alone am responsible for any faults in the final product.

At the Institute for Health Metrics and Evaluation: Tom Achoki, Miriam Alvarado, Charles Atkinson, Ian Bolliger, Dane Boog, Anne Bulchis, Jim Bullard, Kelly Campbell, Emily Carnahan, Brent Christofferson, Michelle Colyar-Cooper, Pete Crow, Joseph Dieleman, Herbert Duber, Laura Dwyer-Lindgren, Linda Ettinger, Abraham Flaxman, Kyle Foreman, Michael Freeman, Emmanuela Gakidou, Diego Gonzalez-Medina, Casey Graves, Michael Hanlon, Gillian Hansen, William Heisel, Spencer James, Nicole Johns, Nicholas Kassebaum, Patricia Kiyono, Katherine Leach-Kemon, Carly Levitz, Stephen Lim, Katherine Lofgren, Rafael Lozano, Michael MacIntyre, Abigail McLain, Ali Mokdad, Kelsey Moore, Kate Muller, Tasha Murphy, Mohsen Naghavi, Summer Ohno, Katrina Ortblad, Jill Oviatt, David Phillips, Kelsey Pierce, Peter Serina, Peter Speyer, Rhonda Stewart, Julie Vithoulkas, Haidong Wang, Sarah Wulf, and Brittany Wurtz, among many others.

At Aga Khan University: Zulfiqar Bhutta. At Brigham and Women's Hospital: Howard Hiatt and Marshall Wolf, among others. At the University of California, San Francisco: Richard Feachem and Jaime Sepulveda. At the University of Cape Town:

George Mensah. At the Dartmouth Institute for Health Policy & Clinical Practice: Andrew Kartunen. At Harvard University: Barry Bloom, Dan Brock, Marah Brown, Panka Deo, Nir Eyal, Julio Frenk, Kelly Friendly, Gary King, Felicia Knaul, Catherine Michaud, Alicair Peltonen, Elizabeth Salazar, Joshua Salomon, Julie Shample, Lawrence Summers, and Daniel Wikler, among others. At Imperial College London: Majid Ezzati. At the London School of Hygiene & Tropical Medicine: Heidi Larson and Peter Piot. At the University of Melbourne: Jed Blore. At the University of Oxford: Richard Peto. At the University of Queensland: Theo Vos. At the University of Tokyo: Kenji Shibuya. At Tsinghua University: Andy Qingan Zhou. At the University of Washington: Brianne Adderley, King Holmes, Dean Jamison, Ruth Mahan, Paul Ramsey, and Judith Wasserheit, among others.

At the Bill & Melinda Gates Foundation: Stefano Bertozzi, Kathy Cahill, Bill Gates, William Gates Sr., Melinda Gates, Mimi Gates, Toni Hoover, Rachel Lonsdale, Trevor Mundel, Jeff Raikes, and Philip Setel. At the China Medical Board: Lincoln Chen. At the Health Effects Institute: Aaron Cohen. At the International Children's Center: Tomris Türmen. At Partners in Health: Emily Bahnsen, Paul Farmer, Jon Niconchuk, and Gretchen Williams, among others. At PATH: Ellen Cole and Steve Davis. At Save the Children: Simon Wright. At the Vitality Group: Francois Millard. At the Washington Global Health Alliance: Lisa Cohen.

At the GAVI Alliance: Seth Berkley and Peter Hansen. At the Global Fund to Fight AIDS, Tuberculosis, and Malaria: Mark Dybul. At the Inter-American Development Bank: Kei Kawabata and Diana Pinto. At the Joint United Nations Programme on HIV/AIDS: Paul De Lay. At the Pan American Health Organization: Marcos Espinal, Fatima Marinho, Maristela Monteiro, and Mirta Roses Periago, among others. At the United Nations

Children's Fund: Mickey Chopra. At the United States Agency for International Development and the Demographic and Health Surveys Program: Jacob Adetunji, John Borrazzo, Trevor Croft, Troy Jacobs, Richard Rinehart, and Shea Rutstein, among others. At the World Bank: Cristian Baeza, Lai-Foong Goh, Jim Yong Kim, Richard Mills, and Marc Shotten, among others. At the World Health Organization: Ties Boerma, Elizabeth Mason, and Colin Mathers, among others. At the WHO Regional Office for Europe: Claudia Stein.

At the Australian Department of Health: Jane Halton. At the Ghana Health Service: Frank Nyonator. At Guinea's National Institute of Public Health: Lamine Koivogui. At the Public Health Foundation of India: K. Srinath Reddy. At Japan's National Institute of Public Health: Tomofumi Sone. At the Office of the Prime Minister of Norway: Tore Godal. At Panama's Gorgas Memorial Institute for Health Studies: Javier Nieto. At the Ugandan Ministry of Health: Christine Ondoa. At the United Kingdom Department of Health and National Health Service: Sally Davies, Adrian Davis, and Stephen King, among others. At the United States Centers for Disease Control and Prevention: Danielle Iuliano and Marc-Alain Widdowson, among others. At the (U.S.) Institute of Medicine: Harvey Fineberg. At Yemen's Ministry of Public Health & Population: Jamal Nasher.

At *The Guardian*: Ian Katz. At *Humanosphere*: Tom Paulson. At *The Lancet*: Daisy Barton, Pam Das, and Richard Horton. At Minerva Strategies: Joy Portella. At ProPublica: Richard Tofel.

At Silver Lake: David Roux. At Trilogy Partnership: John Stanton. At Windhaven Investment Management: Stephen Cucchiaro.

Alan Lopez, Chris Murray, and their friends and family: Thomas Culhane, Lene Mikkelsen, Inez Mikkelsen-Lopez, John Murray, Megan Murray, and Nigel Murray, among others.

Facts and quotations in the text come from these people, from the Global Burden of Disease study, as published in *The Lancet* on December 15, 2012, and made available in greater detail online on March 5, 2013 at *healthdata.org*, and from the following sources, among others. In my research, I used the libraries of Harvard University, the University of Montana, and Northwestern University. In many of my interviews, I used Skype. In my writing, I used Scrivener.

Introduction: Counting Everything When Everything Counts

xi   In 147 of 192 countries: David E. Phillips, Rafael Lozano, Mohsen Naghavi, Charles Atkinson, Diego Gonzalez-Medina, Lene Mikkelsen, Christopher J. L. Murray, and Alan D. Lopez. "A Composite Metric for Assessing Data on Mortality and Causes of Death: The Vital Statistics Performance Index." *Population Health Metrics* 12, no. 14 (May 14, 2014).

xii from $5.8 billion to $29.4 billion a year: Institute for Health Metrics and Evaluation. *Financing Global Health 2013: Transition in an Age of Austerity.* Institute for Health Metrics and Evaluation, 2014: 74–75.

xii 10 percent of the global economy: World Health Organization Global Health Observatory. "Total Expenditure on Health as a Percentage of Gross Domestic Product." Accessed July 26, 2014. http://www.who.int/gho/health_financing/total_expenditure/en/.

Chapter One: Murray, Murray, Murray, and Murray

15   the first paper authored by Murray, Murray, Murray, and Murray: M. John Murray, Nigel J. Murray, Anne B. Murray, and Megan B. Murray. "Refeeding-Malaria and Hyperferraemia." *The Lancet* 305, no. 7908 (March 22, 1975): 653–654.

16   Chris's first official publication: M. John Murray, Anne B. Murray, Megan B. Murray, and Christopher J. Murray. "Somali Food Shelters in the Ogaden Famine and Their Impact on Health." *The Lancet* 307, no. 7972 (June 12, 1976): 1283–1285.

17   "Armchair logic plays little place": M. John Murray, Anne B. Murray, Nigel J. Murray, Megan B. Murray, and Christopher J. Murray. "Reply to Letter by Stephenson and Latham." *The American Journal of Clinical Nutrition* 32, no. 4 (April 1979): 732.

Chapter Two: The Third World and the Nerd World

24   "brilliant and with many potential applications": Edward O. Wilson. Letter

to the Hoopes Prize Committee (May 16, 1984). In Christopher James Livingstone Murray. "Biogeographic Theory and Its Application to the Kenya Rangelands." Undergraduate thesis, Harvard University, 1984.

26   his Oxford doctoral dissertation: Christopher J. L. Murray. "The Determinants of Health Improvement in Developing Countries." Ph.D. thesis, Merton College, University of Oxford, 1988.

26   "There is little doubt": Thomas McKeown. *The Role of Medicine: Dream, Mirage, or Nemesis?* Basil Blackwell, 1979: 85. Quoted in Murray, "The Determinants of Health Improvement in Developing Countries": 1.

27   per capita incomes of, at most, $330: Murray, "The Determinants of Health Improvement in Developing Countries": 5, 9, 14.

27   In Costa Rica, income per capita was $1,020: Ibid., 17.

27   a highly influential 1985 report: Kenneth S. Warren, Julia A. Walsh, and Scott B. Halstead, eds. *Good Health at Low Cost.* Rockefeller Foundation, 1985.

28–29 Murray counted at least five separate models: Christopher J. Murray. "A Critical Review of International Mortality Data." *Social Science & Medicine* 25, no. 7 (July 1987): 773–781.

Chapter Three: How to Die with Statistics

35   Lopez's thesis: Alan D. Lopez. "Which Is the Weaker Sex? A Study of the Differential Mortality of Males and Females in Australia." Ph.D. thesis, the Australian National University, 1978.

43   a scathing paper summarizing his work: Christopher J. Murray. "A Critical Review of International Mortality Data." *Social Science & Medicine* 25, no. 7 (July 1987): 773–781.

44   "they are all considered reliable": United Nations. *Demographic Yearbook 1983.* United Nations, 1985: 97. Quoted in Ibid., 776.

Chapter Four: Missing Persons

50   "Our most striking finding": Commission on Health Research for Development. *Health Research: Essential Link to Equity in Development.* Oxford University Press, 1990: 29.

52   7.1 million people annually: Christopher J. Murray, Karel Styblo, and Annik Rouillon. "Tuberculosis in Developing Countries: Burden, Intervention and Cost." *Bulletin of the International Union Against Tuberculosis and Lung Disease* 65, no. 1 (March 1990): 9–10.

52   "These are the parents": Ibid., 21–22.

53   "We estimate that the total increased cost": Ibid., 22.

53    $4.1 billion before the end of the decade: Barry R. Bloom and Christopher J. Murray. "Tuberculosis: Commentary on a Reemergent Killer." *Science* 257, no. 5073 (August 21, 1992): 1061.

54   Bloom and Murray published an article: Ibid., 1055–1064.

55   more than five million lives: Philippe Glaziou and others. "Lives Saved by Tuberculosis Control and Prospects for Achieving the 2015 Global Target for Reducing Tuberculosis Mortality." *Bulletin of the World Health Organization* 89, no. 8 (published online May 31, 2011, and in print August 2011): 573–582.

55   "Nearly 90 percent of children": Richard G. A. Feachem, Tord Kjellstrom, Christopher J. L. Murray, Mead Over, and Margaret A. Phillips, eds. *The Health of Adults in the Developing World*. Oxford University Press, 1992: 1.

56   all available records circa the year 1985: Alan D. Lopez. "Causes of Death: An Assessment of Global Patterns of Mortality around 1985." *World Health Statistics Quarterly* 43, no. 2 (1990): 91–104.

57   smoking-related causes: Ibid.

60   "Despite a great change": James F. Fries. "Aging, Natural Death, and the Compression of Morbidity." *The New England Journal of Medicine* 303, no. 3 (July 17, 1980): 130.

Chapter Five: The Big Picture

63   a comprehensive review: Dean T. Jamison, W. Henry Mosley, Anthony R. Measham, and José-Luis Bobadilla, eds. *Disease Control Priorities in Developing Countries*. Oxford University Press, 1993.

70   "Because good health increases the economic productivity": World Bank. *World Development Report 1993: Investing in Health*. Oxford University Press, 1993: back cover.

Chapter Six: A Global Checkup

78   Initial Global Burden of Disease Categories: World Bank. *World Development Report 1993: Investing in Health*. Oxford University Press, 1993: 216–219.

79   cause-of-death claims can be garbage: Colin D. Mathers, Doris Ma Fat, Mie Inoue, Chalapati Rao, and Alan D. Lopez. "Counting the Dead and What They Died From: An Assessment of the Global Status of Cause of Death

Data." *Bulletin of the World Health Organization* 83, no. 3 (March 2005): 171–177c.

80   Six percent of very young children: Christopher J. L. Murray and Alan D. Lopez. "Global and Regional Descriptive Epidemiology of Disability: Incidence, Prevalence, Health Expectancies and Years Lived with Disability." In Christopher J. L. Murray and Alan D. Lopez, eds. *The Global Burden of Disease: A Comprehensive Assessment of Mortality and Disability from Diseases, Injuries, and Risk Factors in 1990 and Projected to 2020.* Harvard University Press, 1996: 213.

81   Early Burden of Disease Disability Severity Weighting: Christopher J. L. Murray. "Rethinking DALYs." In Murray and Lopez, eds. *The Global Burden of Disease: A Comprehensive Assessment of Mortality and Disability from Diseases, Injuries, and Risk Factors in 1990 and Projected to 2020*: 40.

83   less than half of total health loss: World Bank, *World Development Report 1993: Investing in Health*: 27.

83   Breaking the data down by sex, age, and region: Ibid., 215–225.

84   "Choices between competing health priorities": Murray and Lopez, "Global and Regional Descriptive Epidemiology of Disability: Incidence, Prevalence, Health Expectancies and Years Lived with Disability": 202.

85   his final assessment of disease burden: Christopher J. L. Murray. "Quantifying the Burden of Disease: The Technical Basis for Disability-Adjusted Life Years." *Bulletin of the World Health Organization* 72, no. 3 (March 1994): 429–445.

85   "doctors' and nurses' time": Sudhir Anand and Kara Hanson. "Disability-Adjusted Life Years: A Critical Review." *Journal of Health Economics* 16, no. 6 (December 1997): 692.

88   introduced on page one: World Bank, *World Development Report 1993: Investing in Health*: 1.

Chapter Seven: Home and Away

90   a report on mental health problems: Robert Desjarlais, Leon Eisenberg, Byron Good, and Arthur Kleinman, eds. *World Mental Health: Problems and Priorities in Low-Income Countries.* Oxford University Press, 1995.

90   "This Report puts the issue of mental health": Ibid., back cover.

91   a Mexico-specific burden-of-disease study: Rafael Lozano, Christopher J. L. Murray, Julio Frenk, and José-Luis Bobadilla. "Burden of Disease Assessment and Health System Reform: Results of a Study in Mexico." *Journal of International Development* 7, no. 3 (May/June 1995): 555–563.

91   the *World Development Report* had specifically suggested: World Bank. *World Development Report 1993: Investing in Health.* Oxford University Press, 1993: iii.

93   clear convergence with the United States: The fertility rate for Mexico comes from Felicia Marie Knaul and others. "The Quest for Universal Health Coverage: Achieving Social Protection for All in Mexico." *The Lancet* 380, no. 9849 (October 6, 2012): 1261. The fertility rate for the United States comes from Joyce A. Martin and others. "Births: Final Data for 2010." *National Vital Statistics Reports* 61, no. 1 (August 28, 2012): 6.

93   "Evidence-Based Health Policy": Christopher J. Murray and Alan D. Lopez. "Evidence-Based Health Policy—Lessons from the Global Burden of Disease Study." *Science* 274, no. 5288 (November 1, 1996): 740–743.

94   the first four *Lancet* papers: Christopher J. L. Murray and Alan D. Lopez. "Mortality by Cause for Eight Regions of the World: Global Burden of Disease Study." *The Lancet* 349, no. 9061 (May 3, 1997): 1269–1276. Christopher J. L. Murray and Alan D. Lopez. "Regional Patterns of Disability-Free Life Expectancy and Disability-Adjusted Life Expectancy: Global Burden of Disease Study." *The Lancet* 349, no. 9062 (May 10, 1997): 1347–1352. Christopher J. L. Murray and Alan D. Lopez. "Global Mortality, Disability, and the Contribution of Risk Factors: Global Burden of Disease Study." *The Lancet* 349, no. 9063 (May 17, 1997): 1436–1442. Christopher J. L. Murray and Alan D. Lopez. "Alternative Projections of Mortality and Disability by Cause 1990–2020: Global Burden of Disease Study." *The Lancet* 349, no. 9064 (May 24, 1997): 1498–1504.

95   "Too often": Gro Harlem Brundtland. Speech on burden-of-disease concept. Hôpitaux Universitaires de Genève (December 15, 1998).

95   "unfocused, even corrupt": Gro Harlem Brundtland. *Madam Prime Minister: A Life in Power and Politics*. Farrar, Straus and Giroux, 2002: 435.

96   "a small revolution": Ibid., 452.

Chapter Eight: Taking on the World

103   "provide an objective assessment": World Health Organization Global Programme on Evidence for Health Policy. "Global Health Leadership Fellows." Public call for applications (November 1998).

106   defined national health system performance: Christopher J. L. Murray and Julio Frenk. "A Framework for Assessing the Performance of Health Systems." *Bulletin of the World Health Organization* 78, no. 6 (June 2000): 717–731.

106   "do not reflect the opinions": World Health Organization. In Christopher J. L. Murray and Alan D. Lopez. "Mortality by Cause for Eight Regions of the World: Global Burden of Disease Study." *The Lancet* 349, no. 9061 (May 3, 1997): 1276.

Chapter Nine: No One's Sick in North Korea

111   *World Health Report*: World Health Organization. *The World Health Report 2000—Health Systems: Improving Performance.* World Health Organization, 2000.

112   "U.S. Spends More": Elizabeth Olson. "U.S. Spends More Than All Others, but Ranks 37 Among 191 Countries." *The New York Times* (June 21, 2000).

112   "'not accurate'": Jothi Jeyasingam. "WHO's Ranking 'Not Accurate.'" *New Straits Times* (September 5, 2000).

112   *Wall Street Journal Europe* commentary: Robert B. Helms. "Sick List: Health Care à la Karl Marx." *The Wall Street Journal Europe* (June 29, 2000).

112   "reducing the size of the public sector": Oswaldo Cruz Foundation, Ministry of Health, Brazil. "Report of the Workshop 'Health Systems Performance—The World Health Report 2000.'" Oswaldo Cruz Foundation (December 14–15, 2000): 3.

113   Murray was quoted: Olson, "U.S. Spends More Than All Others, but Ranks 37 Among 191 Countries."

113   "The material in this report": World Health Organization, *The World Health Report 2000—Health Systems: Improving Performance*: viii.

113   "'Come and help us'": Gro Harlem Brundtland. Presentation of the World Health Report 2000. Foreign Press Association, London (June 21, 2000).

117   disavowing the rankings: Philip Musgrove. "Judging Health Systems: Reflections on WHO's Methods." *The Lancet* 361, no. 9371 (May 24, 2003): 1817–1820.

117   "embarrassed . . . to be associated": Richard Horton. *Second Opinion: Doctors, Diseases and Decisions in Modern Medicine.* Granta Books, 2003: 325.

117   "The objectives of the health systems performance assessment": Sudhir Anand and others. "Report of the Scientific Peer Review Group on Health Systems Performance Assessment." In Christopher J. L. Murray and David B. Evans, eds. *Health Systems Performance Assessment: Debates, Methods and Empiricism.* World Health Organization, 2003: 839.

Chapter Ten: Racing Stripes

124   "terrified that when Mr. Bush left office": Sheryl Gay Stolberg. "In Global Battle on AIDS, Bush Creates Legacy." *The New York Times* (January 5, 2008).

129   total development assistance for health: Institute for Health Metrics and Evaluation. *Financing Global Health 2013: Transition in an Age of Austerity.* Institute for Health Metrics and Evaluation, 2014: 74–75.

129   domestic health care spending: World Health Organization Global Health

Observatory Data Repository. "WHO Region of the Americas: United States of America Statistics Summary (2002–present)." Accessed July 26, 2014. http://apps.who.int/gho/data/?theme=country&vid=20800.

129   worldwide, the domestic average was more than 10 percent: World Health Organization Global Health Observatory. "Total Expenditure on Health as a Percentage of Gross Domestic Product." Accessed July 26, 2014. http://www.who .int/gho/health_financing/total_expenditure/en/.

129   30 cases of malaria per 100,000 people: Christopher J. L. Murray, Alan D. Lopez, and Suwit Wibulpolprasert. "Monitoring Global Health: Time for New Solutions." *British Medical Journal* 329, no. 7474 (November 6, 2004): 1097.

129   "WHO is ill suited": Ibid., 1099.

131   physicians per person: Felicia Marie Knaul and others. "The Quest for Universal Health Coverage: Achieving Social Protection for All in Mexico." *The Lancet* 380, no. 9849 (October 6, 2012): 1259–1279.

131   fewer than 17 per 1,000: Ibid., 1261.

132   In Iran: Mohsen Naghavi and others. "The Burden of Disease and Injury in Iran 2003." *Population Health Metrics* 7, no. 9 (June 15, 2009).

132   In Australia: Stephen J. Begg, Theo Vos, Bridget Barker, Lucy Stanley, and Alan D. Lopez. "Burden of Disease and Injury in Australia in the New Millennium: Measuring Health Loss from Diseases, Injuries and Risk Factors." *The Medical Journal of Australia* 188, no. 1 (August 2008): 36–40. Theo Vos and others. "Assessing Cost-Effectiveness in Prevention (ACE–Prevention): Final Report." University of Queensland and Deakin University (September 2010).

132   In Thailand: Kanitta Bundhamcharoen, Patarapan Odton, Sirinya Phulkerd, and Viroj Tangcharoensathien. "Burden of Disease in Thailand: Changes in Health Gap between 1999 and 2004." *BMC Public Health* 11, no. 53 (January 26, 2011).

132   In nearby Vietnam: Nguyen Thi Trang Nhung and others. "Estimation of Vietnam National Burden of Disease 2008." *Asia-Pacific Journal of Public Health* (published online November 27, 2013).

133   a separate burden study for the country's indigenous population: Theo Vos, Bridget Barker, Stephen Begg, Lucy Stanley, and Alan D. Lopez. *The Burden of Disease and Injury in Aboriginal and Torres Strait Islander Peoples 2003*. The University of Queensland School of Population Health, 2007. Theo Vos, Bridget Barker, Stephen Begg, Lucy Stanley, and Alan D. Lopez. "Burden of Disease and Injury in Aboriginal and Torres Strait Islander Peoples: The Indigenous Health Gap." *International Journal of Epidemiology* 38, no. 2 (April 2009): 470–477.

133   1.7 to 1.9 times the national rate: Vos, Barker, Begg, Stanley, and Lopez, *The Burden of Disease and Injury in Aboriginal and Torres Strait Islander Peoples 2003*: 5.

133   one in three Aboriginal and Torres Strait Islander teenagers: Ibid., vii.

133   close to $900 million: Australian National Preventive Health Agency. *Strategic Plan 2011–2015*. Australian National Preventive Health Agency, 2011: 4.

134   "Evidence that other countries perform better": Christopher J. L. Murray and Julio Frenk. "Ranking 37th—Measuring the Performance of the U.S. Health Care System." *The New England Journal of Medicine* 362, no. 2 (January 14, 2010): 98.

134   an analysis Murray led at Harvard: Christopher J. L. Murray and others. "Eight Americas: Investigating Mortality Disparities across Races, Counties, and Race-Counties in the United States." *PLoS Medicine* 3, no. 9 (September 2006): 1513–1524.

134   "Ten million Americans": Ibid., 1521.

135   "It is when the public": Ibid., 1523.

136   "issues of intrinsic aptitude": Lawrence H. Summers. Remarks at NBER Conference on Diversifying the Science & Engineering Workforce. The National Bureau of Economic Research (January 14, 2005).

137   "The agreement with the university": David Bank. "Oracle's Ellison Gives $115 Million to Harvard Study." *The Wall Street Journal* (June 30, 2005).

137   "When rumors about the Ellison Institute": Richard Horton. "The Ellison Institute: Monitoring Health, Challenging WHO." *The Lancet* 366, no. 9481 (July 16, 2005): 179–181.

139   "The reason I didn't finish my gift": Josephine Moulds. "Oracle Boss 'Lost Confidence.'" *The Daily Telegraph* (June 28, 2006).

Chapter Eleven: Dinner with Bill

144   "'How are you doing on those books?'" Michael Specter. "What Money Can Buy." *The New Yorker* (October 24, 2005): 66.

145   "'That can't be true'": Bill Gates. Interview by Bill Moyers. *NOW with Bill Moyers* (May 9, 2003).

145   "Something like 'an inactivated polio vaccine'": Melinda Gates. Public remarks at an event attended by the author to celebrate the opening of the Bill & Melinda Gates Foundation visitor center. Bill & Melinda Gates Foundation (February 1, 2012).

145   "The whole thing was stunning": Michael Specter, "What Money Can Buy": 65–66.

146   "That is the most amazing fact": Bill Gates, interview by Bill Moyers.

148   "the financial resources of a middle-sized university hospital": Gro Harlem Brundtland. *Madam Prime Minister: A Life in Power and Politics*. Farrar, Straus and Giroux, 2002: 454.

148   "That began our learning journey": Melinda Gates, public remarks at an event attended by the author to celebrate the opening of the Bill & Melinda Gates Foundation visitor center.

149   "largely preventable or inexpensively curable diseases": World Bank. *World Development Report 1993: Investing in Health.* Oxford University Press, 1993: 25.

149   "The metric of success": Matthew Herper. "With Vaccines, Bill Gates Changes the World Again." *Forbes* (November 21, 2011). Accessed July 26, 2014. http://www.forbes.com/sites/matthewherper/2011/11/02/the-second-coming-of-bill-gates/.

149   "Our starting point in deciding where to focus": Bill & Melinda Gates Foundation. "Global Health: Strategy Overview." Bill & Melinda Gates Foundation, 2010: 4.

151   "definite pastiness": Michael Specter, "What Money Can Buy": 65.

Chapter Thirteen: Missionaries and Converts

174   maternal mortality had in fact dropped: Margaret C. Hogan and others. "Maternal Mortality for 181 Countries, 1980–2008: A Systematic Analysis of Progress Towards Millennium Development Goal 5." *The Lancet* 375, no. 9726 (published online April 12, 2010, and in print May 8, 2010): 1609–1623.

175   "The findings, published in the medical journal *The Lancet*, challenge": Denise Grady. "Maternal Deaths Decline Sharply Across the Globe." *The New York Times* (April 13, 2010).

175   "Maternal Deaths Worldwide Drop by Third": World Health Organization, United Nations Children's Fund, United Nations Population Fund, and World Bank. "Maternal Deaths Worldwide Drop by Third." Press release (September 15, 2010).

176   malaria killed twice as many people: Christopher J. L. Murray and others. "Global Malaria Mortality between 1980 and 2010: A Systematic Analysis." *The Lancet* 379, no. 9814 (February 4, 2012): 413–431.

176   "Malaria Deaths Hugely Underestimated": Neil Bowdler. "Malaria Deaths Hugely Underestimated—Lancet Study." BBC News (February 2, 2012).

176   "key [IHME] findings do not seem": World Health Organization Global Malaria Programme. "Malaria: WHO Reaction to IHME Paper in *The Lancet*." Public statement (February 3, 2012).

177   "We have seen a huge increase": Institute for Health Metrics and Evaluation. "Malaria Kills Nearly Twice as Many People Than Previously Thought, but Deaths Are Declining Rapidly." Press release (February 2, 2012).

177   an independent child health expert group: Li Liu and others, for the Child Health Epidemiology Reference Group of the WHO and UNICEF. "Global,

Regional, and National Causes of Child Mortality: An Updated Systematic Analysis for 2010 with Time Trends since 2000." *The Lancet* 379, no. 9832 (published online May 11, 2012, and in print June 9, 2012): 2151–2161.

179   "Scientists agree they need better estimates": Gretchen Vogel. "How Do You Count the Dead?" *Science* 336, no. 6087 (June 15, 2012): 1372–1374.

Chapter Sixteen: London Calling

234   seven of the GBD papers: Haidong Wang and others. "Age-Specific and Sex-Specific Mortality in 187 Countries, 1970–2010: A Systematic Analysis for the Global Burden of Disease Study 2010." *The Lancet* 380, no. 9859 (published online December 13, 2012, and in print December 15, 2012): 2071–2094. Rafael Lozano and others. "Global and Regional Mortality from 235 Causes of Death for 20 Age Groups in 1990 and 2010: A Systematic Analysis for the Global Burden of Disease Study 2010." *The Lancet* 380, no. 9859 (published online December 13, 2012, and in print December 15, 2012): 2095–2128. Joshua A. Salomon and others. "Common Values in Assessing Health Outcomes from Disease and Injury: Disability Weights Measurement Study for the Global Burden of Disease Study 2010." *The Lancet* 380, no. 9859 (published online December 13, 2012, and in print December 15, 2012): 2129–2143. Joshua A. Salomon and others. "Healthy Life Expectancy for 187 Countries, 1990–2010: A Systematic Analysis for the Global Burden Disease Study 2010." *The Lancet* 380, no. 9859 (published online December 13, 2012, and in print December 15, 2012): 2144–2162. Theo Vos and others. "Years Lived with Disability (YLDs) for 1160 Sequelae of 289 Diseases and Injuries 1990–2010: A Systematic Analysis for the Global Burden of Disease Study 2010." *The Lancet* 380, no. 9859 (published online December 13, 2012, and in print December 15, 2012): 2163–2196. Christopher J. L. Murray and others. "Disability-Adjusted Life Years (DALYs) for 291 Diseases and Injuries in 21 Regions, 1990–2010: A Systematic Analysis for the Global Burden of Disease Study 2010." *The Lancet* 380, no. 9859 (published online December 13, 2012, and in print December 15, 2012): 2197–2223. Stephen S. Lim and others. "A Comparative Risk Assessment of Burden of Disease and Injury Attributable to 67 Risk Factors and Risk Factor Clusters in 21 Regions, 1990–2010: A Systematic Analysis for the Global Burden of Disease Study 2010." *The Lancet* 380, no. 9859 (published online December 13, 2012, and in print December 15, 2012): 2224–2260.

234   "a landmark event for this journal": Richard Horton. "GBD 2010: Understanding Disease, Injury, and Risk." *The Lancet* 380, no. 9859 (published online December 13, 2012, and in print December 15, 2012): 2053.

235   "People around the world are living longer": Jane Dreaper. "We Live 'Longer but Sicker' as Chronic Diseases Rise." BBC News (December 13, 2012).

235   "Worldwide, hypertension and tobacco smoking": Larry Husten. "Hypertension and Smoking Top List of Global Risk Factors." Forbes.com. December 13, 2012. http://www.forbes.com/sites/larryhusten/2012/12/13/hypertension-and-smoking-top-list-of-global-risk-factors/.

235   "Deaths from infectious disease are down": "Lifting the Burden." *The Economist* (December 15, 2012).

235   "BLOOD PRESSURE: MILLIONS AT RISK": Jo Willey. "Blood Pressure: Millions at Risk." *Daily Express* (December 14, 2012).

240   "Many have questions": Jon Cohen. "A Controversial Close-up of Humanity's Health." *Science* 338, no. 6113 (December 14, 2012): 1414.

240   "an unprecedented effort": Margaret Chan. "From New Estimates to Better Data." *The Lancet* 380, no. 9859 (published online December 13, 2012, and in print December 15, 2012): 2054.

243   "might be able to sit still": Elizabeth Lowry. "Strong Medicine." *Columns* (December 2007).

Chapter Seventeen: Epic Squared

248   financial share of health aid devoted to them: Institute for Health Metrics and Evaluation. *Financing Global Health 2013: Transition in an Age of Austerity*. Institute for Health Metrics and Evaluation, 2014: 36.

248   the impact of education on health: Emmanuela Gakidou, Krycia Cowling, Rafael Lozano, and Christopher J. L. Murray. "Increased Educational Attainment and Its Effect on Child Mortality in 175 Countries between 1970 and 2009: A Systematic Analysis." *The Lancet* 376, no. 9745 (September 18, 2010): 959–974.

250   twenty times the amount of outside aid: Institute for Health Metrics and Evaluation, *Financing Global Health 2013: Transition in an Age of Austerity*: 61.

Chapter Eighteen: From Galileo to Chris Murray

259   annual budget several times the total amount of health aid: U.S. Veterans Health Administration. "VA 2015 Budget Request Fast Facts." Accessed July 26, 2014. http://www.va.gov/budget/docs/summary/Fy2015-FastFactsVAsBudgetHighlights.pdf.

260   Australia spends more money: Australian Institute of Health and Welfare. "Expenditure FAQ." Accessed July 26, 2014. http://www.aihw.gov.au/expenditure-faq/.

261 "Most media, Australia a notable exception": Tom Paulson. "Ten Disease Burden Stories from Around the Globe." *Humanosphere* (March 7, 2013). http://www.humanosphere.org/global-health/2013/03/around-the-world-with-the-global-burden-of-disease/.

265 "A business has increasing profit": Bill & Melinda Gates Foundation. "Annual Letter from Bill Gates." Bill & Melinda Gates Foundation, 2013: 2–3.

266 $2.5 billion to expand vaccine access and development: Bill & Melinda Gates Foundation. "Vaccine Delivery: Strategy Overview." Accessed July 26, 2014. http://www.gatesfoundation.org/What-We-Do/Global-Development/Vaccine-Delivery.

267 a detailed analysis specific to the United Kingdom: Christopher J. L. Murray and others. "UK Health Performance: Findings of the Global Burden of Disease Study 2010." *The Lancet* 381, no. 9871 (March 5, 2013): 997–1020.

268 "I want us to be up there with the best": UK Department of Health. "Living Well for Longer." UK Department of Health (March 5, 2013): 3.

269 "The State of US Health": US Burden of Disease Collaborators. "The State of US Health, 1990–2010: Burden of Diseases, Injuries, and Risk Factors." *Journal of the American Medical Association* 310, no. 6 (published online July 10, 2013, and in print August 14, 2013): 591–606.

270 "consistent with and incorporating UN agency": World Health Organization Health Statistics and Information Systems. "Global Burden of Disease." Accessed July 26, 2014. http://www.who.int/healthinfo/global_burden_disease/gbd/en/.

271 "individuals or groups that have used Global Burden of Disease": Institute for Health Metrics and Evaluation. "IHME Launches Roux Prize to Reward Use of Global Burden of Disease Evidence to Improve Health." Press release (November 13, 2013).

273 Murray's salary was $488,000: *The Spokesman-Review*. "Washington State Employee Salary Database: 2012 salaries." Accessed July 26, 2014. http://data.spokesman.com/salaries/state/2013/all-employees/.

Afterword: How to Live a Longer and Healthier Life According to the Global Burden of Disease

286 Top 10 U.S. Counties in Terms of Change in Life Expectancy: Institute for Health Metrics and Evaluation. *The State of US Health: Innovations, Insights, and Recommendations from the Global Burden of Disease Study*. Institute for Health Metrics and Evaluation, 2013: 39.

287 Avoiding Early Mortality—Education Improves the Odds: Institute for Health Metrics and Evaluation. "Adult Mortality Rates by Country and Sex." Accessed April 18, 2013. http://www.healthmetricsandevaluation.org/tools/

data-visualization/adult-mortality-rates-country-and-sex-global-1970-2010.

288 One theory explains Japan's success: Christopher J. L. Murray. "Why Is Japanese Life Expectancy So High?" *The Lancet* 378, no. 9797 (September 24, 2011): 1124–1125.

290 Top 10 Countries for Overall Health System Performance: World Health Organization. *The World Health Report 2000—Health Systems: Improving Performance.* World Health Organization, 2000: 200.

290 Top 10 Countries in Terms of Health Spending per Capita: World Bank. "Health Expenditure Per Capita." Accessed July 26, 2014. http://data.world bank.org/indicator/SH.XPD.PCAP?order=wbapi_data_value_2000+w bapi_data_value&sort=desc&page=2.

data discrepancies, 105, 205
and education, 248
improvements in, 195–96, 264
in Mexico, 92, 131
and Millennium Development
  Goals, 172–73, 178
UNICEF's, 219
children
  breast-feeding, 164, 165, 206,
    208–9
  and injuries from ages five to
    fourteen, 196
  and Malthus's theory, 146
  preterm birth complications,
    204
  success of Millennium Develop-
    ment Goals, 206
China
  cause of death in, 261
  COPD, 126
  Global Burden launch, 246
  health data regulations, 159
  health loss per person, 265
  life expectancy statistics, 27,
    195
  tuberculosis treatment funding,
    54–55
  World Health Report ranking,
    112
cholera, 33, 182
Church World Service, 7
cirrhosis, 12, 223
collaboration on burden-of-disease
  analysis, 267–68
Commission on Health Research
  for Development, 49–51
communicable diseases, 48, 77,
  82–83, 92–93. See also
  specific diseases
Comoros, 16
compression of morbidity, 60, 199
COPD (chronic obstructive pulmonary disease), 58, 201–2,
  204, 206–7, 283
Costa Rica, 27, 215

Culhane, Thomas Henry Rassam,
  19–25

DALYs (disability-adjusted life
    years)
  about, 65–69, 86–87
  dynamic visuals, 264
  and Gates, 148, 149
  Global Burden statistics, 203–4
  scientific papers on, 167
  usage of data, 90–93
  See also entries beginning with
    "Global Burden of Disease
    study"
Dartmouth Institute for Health
    Policy & Clinical Practice,
    254
data
  about, xi, 134–35, 266
  and cause of illness and death,
    xi–xiv
  contradictions in, 28–29,
    37–41, 43–44, 105, 171–72,
    174–75, 177–79
  current usage, xv–xvi
  disease classification differences
    across countries, 77–79
  and epidemics, 33–34
  faulty data, 44, 128–29, 161–62
  and granularity, 75
  on life expectancy and mortal-
    ity rate, 27–30, 31
  processes for gathering, 37–41,
    107–8, 158–63
  and tuberculosis treatment,
    51–53
  weighting, 81, 85, 133, 167,
    168–69
  WHO and UN data on mor-
    tality rate in poor countries,
    37–41, 105
  World Bank traffic injuries
    database, 128
  See also DALYs; dynamic data
    visualization

Guatemala, 215, 216

Haiti, 59
Halton, Jane, 133, 192
Hanson, Kara, 85
Harvard, 19–23, 24–25
Harvard Center for Population and
    Development Studies (Pop
    Center), 48–51, 53–55, 76,
    84–86, 89–90, 95
Harvard Initiative for Global
    Health (HIGH), 125–26,
    127–29, 135–40
Harvard Medical School, 47, 48
Harvard School of Public Health,
    89, 94–95
health
    investment leading to economic
        productivity, 70–71
    spending worldwide, xii, 50–51
    worldwide statistics shortage,
        28–29
    YLDs, 65–69, 80, 158, 199–
        205, 284
    YLLs, 65–69, 81, 202, 227–28
    See also adult health; children;
        DALYs; Global Burden
        guide to living a healthier
        life
"Health and the Economy" (Na-
    tional Institute of Public
    Health of Mexico), 91
health insurance, 27, 130–31, 283,
    288
health measurement, 137, 138,
    152–53, 171, 191. See also
    entries beginning with
    "Global Burden of Disease
    study"
health spending policy decisions
    and appearance of health de-
        partment, 252
    and contradictory data, 28–29,
        37–41, 43–44, 105, 171–72
    and DALYs, 90–93

data and competitive environ-
    ment, 174
evaluating coverage of different
    interventions, 251
and faulty data, 44, 128–29
and Global Burden study, 239
importance of data, 179–80
paths to influencing, 221–22
prevention focus, 165–66
rankings of spending and per-
    formance by country, 290
review of disease control priori-
    ties in developing countries,
    63–64, 69–71
See also Millennium Develop-
    ment Goals
health systems
    inequalities across countries,
        250–51
    performance, 251
    rankings of spending and per-
        formance by country, 290
    and shifting age of population,
        59, 195–96, 198
    Yemen's internal health system
        assessment, 255–56
    See also health spending policy
        decisions; World Health
        Report
healthy life expectancy, 262,
    288–89. See also Global
    Burden guide to living a
    healthier life
heart disease. See IHD
hepatitis A, 11
Hernández-Ávila, Mauricio, 222
high blood pressure (hypertension),
    206, 208, 270
high blood sugar, 207. See also
    diabetes mellitus
HIV/AIDS
    Global Burden statistics,
        203–4, 205
    international health institution
        for, 96

combining epidemiology and demography, 39–41

compiling data on morality and causes of death, 37–41

on countries' responses to *World Health Report*, 114–15

on global burden in early 1990s, 246

Global Burden of Disease subgroup in Australia, 270–71

and IHME, 158, 216–18

on importance of knowing cause of death, 177

leaving WHO, 120

in London, 237

on morbidity and disability, 68

and Murray, 42, 180–81

sex differences in mortality studies, 34–36

on statistics, 34

and tobacco, 57

and WHO Evidence and Information for Policy cluster, 99, 108

on WHO's 2000–2011 global burden-of-disease estimates, 270

on WHO's ineffective global monitoring, 129

See also entries beginning with "Global Burden of Disease study"

Lozano, Rafael, 91–92, 157–58, 192, 214–16, 236

lymphatic filariasis, 231

MacIntyre, Michael, 155

malaria

as cause of death, 40

faulty data on prevalence of, 129

Global Burden statistics, 203–4

IHME exposes underestimation of deaths from, 176–77

and iron in supplements, 13–15

and Millennium Development Goals, 172, 176–77, 179

Malaysia, 112, 158

malnutrition, 13–15, 16, 26–27, 39

Malthus, Thomas, 146

"Maternal Deaths Worldwide Drop by Third" *(Trends in Maternal Mortality)*, 175

maternal mortality, 120, 172, 173, 174–76, 272

Mathers, Colin, 217

Mauritius, 158

McKeown, Thomas, 26–27

measles, 40, 64

measure of national health systems' performance. *See World Health Report*

measures of health, 137, 138, 152–53, 171, 191. *See also entries beginning with* "Global Burden of Disease study"

media coverage, importance of, 220–21

mental health problems

and burden-of-disease analysis, 132

in Costa Rica, 215

depression, 132, 200, 201–4, 216

GBDx dynamic display of, 228–30

and WHO, 97

Mexico, 90–93, 130–31, 220–21, 287

Michaud, Catherine, 90, 126, 128, 143, 275

Middle East, 23–24, 216–18. *See also specific countries*

middle-income countries, 57, 134, 197

migraines, 81, 201–3, 284

Millennium Development Goals

about, 120, 172–73

and child mortality rate, 172–73, 178

and HIV/AIDS, 172, 205

improvement, 27–30, 31
Speyer, Peter, 159–60, 235–36
Sri Lanka, 27
starvation, 13–15, 16, 26–27, 39
"State of US Health, 1990–2010"
    (*Journal of the American Medical Association*),
    269–70
statistics, 34, 35–41, 39–41. *See also* data
Stonesifer, Patty, 151
stroke, 57, 203–4, 206–7
Styblo, Karel, 52–53, 55
Sudan, 217
Summers, Larry, 70, 71, 124–25,
    136, 139
survival advantage of women,
    200–201
Switzerland, 287, 288, 289

10/90 gap, 50–51
Thailand, 132, 158
tobacco smoking, 34, 57, 165–66,
    206, 270, 283
Tolkien, J. R. R., 26
traffic accidents
    in developing countries, 83
    as Global Burden category, 78
    Global Burden statistics, 204
    health consequences, 79–80
    and helmet requirement in
        Vietnam, 132–33
    insurance coverage for emergency care, 130
    in Iran, 132
    World Bank traffic injuries
        database, 128
    years of healthy life lost, 92,
        132, 183
Trends in Maternal Mortality, 175
tuberculosis (TB), 51–54, 123–24,
    172

United Kingdom burden-of-disease
    analysis, 267–68

United Nations
    about, 30
    assumption of steady life expectancy increase, 42–45
    and IHME, 187
    Murray's research at, 42
    and WHO Evidence and Information for Police cluster,
        105
    See also Millennium Development Goals
United Nations Children's Fund
    (UNICEF)
    child mortality rate, 105, 187,
        219
    data from, 43, 177–78, 187
    Horton's criticism of, 240
    and IHME findings, 173,
        175–76, 210, 219
United Nations Conference on
    Sustainable Development,
    Rio Plus 20, 210
United Nations Population Division, 29, 38–39, 173, 187
United Nations Programme on
    HIV/AIDS (UNAIDS),
    96, 210
United States
    Global Burden statistics, 227,
        281–83
    and Global Burden study,
        269–70
    health differences between
        different populations, 134
    health loss per person, 265
    income level and lousy outcomes, 252–53
    life expectancy figures, 261–62
    Mexico compared to, 130
    ranking in *World Health Report*,
        112–13
    resistance to unified health
        system, 133–34
    states selling hospital inpatient
        data, 159

ABOUT THE AUTHOR

Jeremy N. Smith has written for *Discover*, the *Christian Science Monitor*, and the *Chicago Tribune*, among many other publications. His first book, *Growing a Garden City*, was one of *Booklist*'s top ten books on the environment for 2011. Born and raised in Evanston, Illinois, he is a graduate of Harvard College and the University of Montana. He lives in Missoula, Montana, with his wife and young daughter.